Synthese Library

Studies in Epistemology, Logic, Methodology, and Philosophy of Science

Volume 514

The aim of *Synthese Library* is to provide a forum for the best current work in the methodology and philosophy of science and in epistemology, all broadly understood. A wide variety of different approaches have traditionally been represented in the Library, and every effort is made to maintain this variety, not for its own sake, but because we believe that there are many fruitful and illuminating approaches to the philosophy of science and related disciplines.

Special attention is paid to methodological studies which illustrate the interplay of empirical and philosophical viewpoints and to contributions to the formal (logical, set-theoretical, mathematical, information-theoretical, decision-theoretical, etc.) methodology of empirical sciences. Likewise, the applications of logical methods to epistemology as well as philosophically and methodologically relevant studies in logic are strongly encouraged. The emphasis on logic will be tempered by interest in the psychological, historical, and sociological aspects of science. In addition to monographs *Synthese Library* publishes thematically unified anthologies and edited volumes with a well-defined topical focus inside the aim and scope of the book series. The contributions in the volumes are expected to be focused and structurally organized in accordance with the central theme(s), and should be tied together by an extensive editorial introduction or set of introductions if the volume is divided into parts. An extensive bibliography and index are mandatory.

Manuel Gustavo Isaac • Steffen Koch
Kevin Scharp
Editors

New Perspectives on Conceptual Engineering - Volume 2

Across Philosophy

Springer

Editors
Manuel Gustavo Isaac
GESDA - Geneva Science and Diplomacy Anticipator
Geneva, Switzerland

Steffen Koch
Department of Philosophy
Bielefeld University
Bielefeld, Germany

Kevin Scharp
University of Illinois Urbana-Champaign
Champaign, IL, USA

ISSN 0166-6991 ISSN 2542-8292 (electronic)
Synthese Library
ISBN 978-3-031-98523-2 ISBN 978-3-031-98524-9 (eBook)
https://doi.org/10.1007/978-3-031-98524-9

This Springer imprint is published by the registered company Springer Nature Switzerland AG
The registered company address is: Gewerbestrasse 11, 6330 Cham, Switzerland

This collection of essays draws on lectures presented in the Conceptual Engineering Webinar Series. The webinar series was launched by Manuel Gustavo Isaac and Kevin Scharp in March 2020 at the University of St Andrews before being subsequently co-hosted with the University of Zurich. In 2022, it became an independent initiative run by the Conceptual Engineering Network (https://www.conceptualengineering.xyz), whose coordinators include Steffen Koch. The Conceptual Engineering Network curates the Conceptual Engineering YouTube channel (https://www.youtube.com/@ConceptualEngineering) for online dissemination.

Contents

Contributors

Kwame Anthony Appiah New York University, New York City, NY, USA

Sanford C. Goldberg Department of Philosophy, Northwestern University, Evanston, IL, USA

Arché Research Center, University of St Andrews, St Andrews, UK

Frank Jackson Australian National University, Canberra, Australia

Teresa Marques LOGOS Research Group, BIAP – Barcelona Institute of Analytic Philosophy, University of Barcelona, Barcelona, Spain

Tristram McPherson Ohio State University, Columbus, OH, USA

Mari Mikkola University of Amsterdam, Amsterdam, Netherlands

Jennifer Nado University of Hong Kong, Hong Kong, China

David Plunkett Dartmouth College, Hanover, NH, USA

Mona Simion Cogito Epistemology Research Centre, University of Glasgow, Glasgow, Scotland, UK

Introduction

Conceptual engineering is a tradition in contemporary philosophy that focuses on evaluating the things we use to represent the world, like our concepts and the meanings of our words. Moreover, conceptual engineering is also about what to do when these representational devices turn out to be defective or problematic. Should they be replaced? With what? When is it reasonable or permissible or obligatory to continue using a problematic concept or word? And finally, what kinds of philosophical problems or issues can reasonably be addressed by using the techniques of conceptual engineering? What are the major obstacles and challenges for conceptual engineering as a philosophical methodology? Although philosophers in the Western tradition have long been engaged in the kinds of questions addressed by conceptual engineering, there has recently been an explosion of work explicitly on its methodology, case studies, and further applications. Thus, the issues explored in the recent literature on conceptual engineering mark a distinctive new phase in philosophical reflection on these topics.

Although there are plenty of historical examples of Western philosophers evaluating various philosophical concepts and making proposals for how to improve them, the work done on conceptual engineering as an explicit methodology is familiar from Rudolph Carnap's work on the method of explication and his attention to explications of truth, confirmation, and probability in the mid-twentieth century. P. F. Strawson's famous criticism of explication as a pointless changing of the topic is a reaction to Carnap's work of the same era. The most recent upsurge in work on conceptual engineering is inspired by theorizing about the concept of truth and on concepts of race and gender, especially in writing by Sally Haslanger who advocates a method she calls *amelioration*. The most recent surge in interest in conceptual engineering came about when philosophers working in each of these camps began to reflect on the methods and examples of conceptual engineering together. Now, a mere 10–15 years later, there is a thriving literature on dozens of aspects of the method of conceptual engineering, there are too many applications of the method to specific cases across philosophy to count, and there are important interdisciplinary connections to many other areas of study including physics, biology, computer science, artificial intelligence, psychology, sociology, economics, finance, and more. In addition, theorists are reflecting on the relationships between foundational issues in the general methodology, the paradigmatic case studies across philosophy, and applications of the method to a wide variety of topics in related areas.

This is volume two of a three-volume set on conceptual engineering. The first volume—*Foundational Issues*—addresses subjects associated with the general method of conceptual engineering, this volume—*Across Philosophy*—covers cases of conceptual engineering across distinct areas of philosophical theorizing, and the third—*Applied Conceptual Engineering*—is about conceptual engineering applied to topics outwith those that traditionally occupy professional philosophers.

In his contribution "Conceptual Engineering in Context," Frank Jackson provides a very rich account of the role that conceptual engineering plays in philosophical theorizing. His starting point is the observation that conceptual engineering can be described as the identification and labeling of theoretically fruitful properties. Contrary to some recent criticisms of conceptual engineering, Jackson argues, this perspective shows that conceptual engineering is an integral and important part of philosophy. Nonetheless, according to Jackson the explicit recasting of familiar philosophical debates in analytical philosophy as conceptual engineering (e.g., knowledge, Swampman, narrow content, and proper names) sheds new light on what is at stake in these debates and constrains the ways in which they can be advanced. Finally, Jackson considers the relationship between conceptual engineering and conceptual analysis, arguing that the two are not in conflict but can complement each other in various ways.

Jennifer Nado, in "Conceptual Engineering, the Value of Knowledge, and the Value of Understanding," insightfully applies the methodological principles of conceptual engineering to disentangle a paradigm shift debated in contemporary analytic epistemology. Nado starts by characterizing her preferred construal of conceptual engineering as "practical." By Nado's account, conceptual engineering "focuses on the functions or roles that our concepts play in our cognitive lives, and frames conceptual improvement in terms of an increased ability to effectively serve those roles". Nado next applies her role-driven version of conceptual engineering to critically address the "Refocusing Thesis," which calls for re-orienting the focus of epistemology from knowledge to understanding. She then analyzes the epistemic roles of the knowledge concept and of the understanding concept and comparatively assesses their respective merits for the purposes of epistemological theorizing. Nado concludes that, although the concept of understanding is of theoretical interest in contexts with high epistemic standards, the concept of knowledge (or its potential successor) still better fits many of the most relevant epistemic roles for our epistemological theories—thereby endorsing a weak version of refocusing.

The chapter, "Engineering Evidence," by Mona Simion, applies conceptual engineering to the concept of evidence in epistemology. What we care about both within and outwith philosophy, Simion recalls, is not just the nature of evidence, but "what it is for a subject to *have* evidence". Simion next diagnoses a tension between the lay and the philosophical conceptions of the having relation—the latter "having to do with the limits of one's skull", while the former concerns the easy availability of a piece of evidence. Simion then establishes the theoretical failure of some prominent externalist accounts (namely, reliabilism and knowledge-first) to accommodate evidence and defeat resistance and thus reliably track the corresponding instances in the world; thereby, Simion claims, these accounts fail to play their practical function of predicting accountability. Simion proposes to fix both these theoretical and practical shortcomings of our philosophical conception of the having relation by aligning it with its lay conception, and then re-engineers the lay conception of the having relation as that of being in a position to know. Eventually, Simion ventures, "if expert accounts have causal power over meanings", this could lead to a change in our concepts—a full-blooded conceptual engineering achievement.

Sandford Goldberg's chapter is entitled "Engineering a Concept of Epistemic Justification," and that is exactly what he delivers. Goldberg begins by arguing that there are clear standards for a concept of epistemic justification and we can see that a new concept of epistemic justification does a better job at satisfying them than the one we currently find in the epistemology literature. Goldberg's six criteria include a core epistemic condition on knowledge along with conditions for intellectual responsibility, moral responsibility, legal responsibility, the ethics of belief, and testimonial credibility. He argues as well that the topic of epistemic responsibility is not especially prone to problems like changing the subject or implementation, which do haunt other conceptual engineering proposals. Goldberg's proposal combines elements from evidentialist views and from responsibilist views of justification into a single concept. He proposes the following definition: For all subjects S, propositions p, and evidence E, where E constitutes S's total evidence at time t, S's belief that p is *propositionally justified* at t iff (a) E provides adequate support to S's belief that p and (b) E continues to provide adequate support to S's belief that p even when E is supplemented by the evidence S should have had. Goldberg concludes by arguing that this concept meets the standards he set out better than any of the alternatives.

In their contribution "Conceptual Ethics and The Categories of 'Ideal Theory' and 'Non-Ideal Theory' in Political Philosophy: A Proposal for Abandonment," David Plunkett and Tristram McPherson critically reflect on the use of the "ideal" vs. "non-ideal" distinction in normative political philosophy. Building on an analysis of the various dimensions at work in idealizations in political philosophy, Plunkett and McPherson conclude to the epistemic uselessness and counterproductiveness the "ideal" vs. "non-ideal" distinction in political philosophy. Consequently, the authors propose to abandon use of "(non-)ideal theory" in political philosophy and replace it by focusing "on the particular idealizations that are being made in a given argument, theory, etc.—or, somewhat more broadly, the particular kinds of idealizations". Plunkett and McPherson end up comparing their eliminativist proposal with two alternative options in conceptual ethics, namely, retention and reform.

Teresa Marques in her chapter, "Representing or Shaping Reality? What *Class* Can Teach about *Woman*," addresses the issue of what counts as a legitimate basis for re-engineering concepts? In particular, how are we to balance epistemic considerations against political considerations? According to Mona Simion' epistemic limiting procedure (Simion, 2018), we ought to engineer a concept only if we have all things considered reason to do so, yet this would not result in an epistemic loss. Contra Simion, Paul-Mikhail Podosky (2019) and Robin McKenna (2019) independently argue that ideological concepts create rather than represent social reality, which is why engineering them cannot be done without resulting in a temporary epistemic loss. Marques focuses on Sally Haslanger's ameliorative analyses of gender concepts to articulate objections against both sides of the debate. Drawing on an important analogy between gender concepts and the Marxist analysis of class, she argues, contra Simion, that we have every reason to believe that Haslanger's hierarchical gender concepts are representationally accurate; pace Podosky and McKenna, she argues for a different and more nuanced understanding of amelioration and societal change. Whereas Podosoky and McKenna hold that the goal of an ameliorative analysis is to provide concepts that shape reality so that they become more accurate, Marques argues for the opposite conclusion: both Marxist concepts of class and Haslangerian gender concepts are formulated in the hope of accurately predicting types of unfair equality that will ultimately vanish.

Mari Mikkola also focuses on whether conceptual engineering ought not to result in epistemically less accurate concepts in her chapter entitled "Engineering Ideologically Defective Concepts." According to Mona Simion, it is never legitimate to engineer a concept where this results in an epistemic loss. Mikkola argues, following Paul-Mikhail Podosoky and pace Simion, that we often have good reason to engineer ideologically defective concepts, even if this results in an epistemic loss. However, the author also takes issue with Podosoky's suggested alternative: that engineering is permissible only if we have reason to believe that the proposed concept will make itself accurate in the near future. According to Mikkola, Podosky's optimism about an engineered concept to shape reality in a way so that it becomes accurate is based on a simplistic and ultimately false account of what an ideology is. Once we acknowledge the actual complexity of the matter, Mikkola argues, we should be far less optimistic about the prospects of engineering ideologically defective concepts. But even if conceptual re-engineering of such concepts isn't feasible, we might still engage in constructing altogether new concepts from scratch and in using conceptual ethics to assess the alternatives.

In his contribution, "Conceptual Engineering: Rethinking 'Race,'" Kwame Anthony Appiah sheds new light on some of the major contributions to the understanding of race in Anglophone philosophy over the past half-century. Before turning to race, however, Appiah begins with an insightful general account of engineering, according to which it crucially involves a design specification, an analysis of what it takes to meet that specification, the recognition that any given design specification can be realized in multiple ways and that conceptual innovation is an unpredictable process that works best on a trial-and-error basis. When concepts are engineered, the engineering process should be understood as a movement for social change. And because concepts are embedded in often idealized theories, they can hardly be engineered individually, but must be considered in clusters, and the engineer must often look beyond considerations of truth. Appiah then applies these insights to a thorough reinterpretation of debates about the nature of race. Re-interpreting parts of the debate as one about whether and to what end the concept of race is a concept we should either use, revise, or abandon, he argues, allows for a more nuanced assessment of the options and arguments put forward—including his own.

There is an important connection between philosophical work on the method of conceptual engineering in general and the specific cases to which it has been applied. The chapters in this volume focus on improving specific concepts or groups of concepts from across philosophy. These case studies help us appreciate how the general methodology functions in specific situations and provide paradigms that can be used as standards against which future studies are measured, including those that focus on issues outside the traditional bounds of philosophy.

University of Illinois Urbana-Champaign
Urbana-Champaign, IL, USA
Kevin Scharp

Geneva Science and Diplomacy Anticipator
Geneva, Switzerland
Manuel Gustavo Isaac

Department of Philosophy, Bielefeld University
Bielefeld, Germany
Steffen Koch

Chapter 1
Conceptual Engineering in Context

Frank Jackson

Abstract I argue that conceptual engineering, on the understanding I favour, is the identification and naming of the properties we need for this or that worthwhile purpose, independently of whether they are properties named in extant theories. So understood, it is obvious that conceptual engineering should be supported, and I give some non-controversial examples. I discuss the connection between our description of conceptual engineering in terms of properties and words, and the more usual one in terms of concepts. The implications of conceptual engineering for a number of live issues in analytical philosophy—knowledge, Swampman, narrow content, proper names—are discussed. I conclude by noting that our defence of conceptual engineering does not imply that traditional conceptual analysis has no worthwhile role to play but does imply that its role needs a certain amount of rethinking. I illustrate this with the example of personal identity.

Keywords Conceptual analysis · Narrow content · Conceptual engineering · Knowledge · Personal identity · Swampman · Names

1.1 Some Examples to Set the Scene

(i) Once upon a time, mathematicians counted one as a prime number on the ground that it is a positive integer exactly divisible by itself and one alone. Later, the definition of a prime number was changed to that of a positive integer exactly divisible by only two integers. The concept was rejigged in a way that excluded one from the class of prime numbers.[1] (ii) It is obvious that one finite set can be bigger than

[1] This is a bit of 'philosophers history'. Current mathematicians are explicit that one is not a prime; the extent to which, back in the day, the number one was explicitly included in the primes, as opposed to being included by implication is, I understand, a moot point.

F. Jackson (✉)
Australian National University, Canberra, ACT, Australia
e-mail: frank.jackson@anu.edu.au

M. G. Isaac et al. (eds.), *New Perspectives on Conceptual Engineering - Volume 2*, Synthese Library 514, https://doi.org/10.1007/978-3-031-98524-9_1

another finite set, but can *infinite* sets differ in how big they are? Cantor resolved the matter by specifying what it takes for two sets to be the same size in terms of one-one correspondence and showing that, in terms of that concept, infinite sets can fail to be the same size, and that, in particular, the set of reals is bigger than the set of natural numbers. (iii) Anthropologists decided to make the concept of a sibling precise as follows: x and y are siblings if and only if they have a parent in common. (iv) Philosophers theorising about phenomenal indistinguishability came to realise that the key concept should not be specified as: x and y are indistinguishable in respect to (phenomenal) property P if and only if they cannot be told apart in respect to P, but instead as: x and y are indistinguishable in respect to P if and only if, for all z, x is indistinguishable from z in respect to P if and only if y is indistinguishable from z in respect to P.[2]

I am sure readers can think of other examples. Conceptual engineering in the sense of rejigging or making more precise existing concepts, and in the sense of introducing new concepts, is a commonplace, and is often a good thing. The examples of the preceding paragraph are examples of sensible, warranted bits of conceptual engineering. (I am not suggesting that there are no examples of misguided exercises in conceptual engineering and imagine that many readers have their favourite examples of same.)

If it is a commonplace that conceptual engineering is a good thing, one might well wonder what this could tell us of interest. It is hardly an exciting discovery that calls for analytical philosophers to rethink the way they do business. I think that this natural response—which is, I take it, one theme in Deutsch (2020)—would be a mistake. I think that reflecting on the fact that conceptual engineering is often a good thing, and why this is so, teaches us something important. This paper spells this out. But first we need to note an unfortunate feature of the way I described our examples.

1.2 Another Way to Describe our Examples

The examples were described employing the language of concepts. Philosophers often talk in terms of concepts, and it is hard to introduce a discussion of *conceptual* engineering without using the term or one of its cognates. What is more, I am sure that readers had a sense of what was being said when I talked above in terms of concepts. All the same, there is a problem: the term 'concept' (and its cognates) is a contested term of art that gets used in importantly different ways by different authors. We can, however, describe the examples we used to introduce our subject and make it plausible that conceptual engineering is a good thing without using the term or its cognates.

[2] They were influenced by Goodman (1977, p. 196).

Our examples are examples of engineering the *properties* certain *words* pick out. We spoke of mathematicians rejigging the concept of a prime number but could instead have talked of their changing the way they use 'prime number' in a way that means it refers to a different property. We could have said that what Cantor did was to identify a property that had previously escaped notice, at least as it applies to infinite sets, that of being in one-one correspondence, and to use 'is the same size as' (or the German equivalent) for the property. We could have said that anthropologists decided they needed a term for the property of having a parent in common and recruited 'sibling' for the task. Finally, we could have said that what philosophers working on phenomenal indistinguishability came to realise was which property it was best to use 'is indistinguishable with respect to P' for. We can now restate our earlier observation that it is a commonplace that conceptual engineering is sometimes a good thing. We all agree that it is a good thing to identify which properties are important, and to coin new words, or recruit existing ones, so that we can talk and write about these properties, and, thereby, frame and convey our thoughts about these properties.

Why is it better to put the issue on the table in terms of properties and words rather than in terms of concepts? Although the metaphysics of properties is controversial, we have no choice but to allow that we can, in one way or another, make sense of the *strength* of a *magnetic* field, the *amount* of *spin* Barty puts on her backhand, how *far away* the Sun is, and so on. We must, that is, allow that we can make sense of and think about the general side of the famous particular–general divide. For properties—items in the general side of the ledger—are central to our explanations of how things are as they are, and our predictions. Epidemiologists explain the spread of Covid in terms of its *transmissibility*; physicists explain the behaviour of compasses in terms of the *strength* and *direction* of the Earth's magnetic field; we predict where an arrow will land based on its *direction* and *velocity*; and so on. Furthermore, we had better be able to talk and write about properties; we need, that is, words for the properties. Framing the discussion in terms of properties and words is framing it in terms we all have to admit.

Now that we have explained how our issue can be stated in terms of properties and words, we can spell out *a* way to think of the 'concept' way of describing the examples that we started with. We can think of concepts as matching up with the properties words are used to pick out. When Cantor gave us a new concept, he gave us a new property to use 'is the same size as' for, a property that applies in the case of infinite sets. The change in the concept of being a prime number corresponds to a change in the property mathematicians pick out when they use 'is a prime number'. When philosophers of perception came to realise that the concept they needed wasn't x and y being pairwise indistinguishable, but instead x being indistinguishable with respect to z if and only if y is indistinguishable with respect to z, for all z, they came to realise the best property to use 'is indistinguishable' for. And so on. We have, that is, a way to move between the two ways of saying things. We will take advantage of this at places in what follows.

1.3 A Sense of What Is to Come

I said earlier that I would spell out how reflecting on the fact that conceptual engineering is often a good thing, and why it is, teaches us something important. I can now say what the reflecting amounts to: it is reflecting on which properties are important for one or another worthwhile purpose, and which words we might use for these properties. Or, to say it in the language of concepts, it is reflecting on which concepts are important for one or another purpose. I am going to survey some familiar issues in philosophy and show how they look when approached in this spirit.

To give a flavour of what is to come, think back on the examples of conceptual engineering we started with. For each, there is a clear justification for focussing on the property in question and giving it a name. In the prime number example, it is to make the Factorisation Theorem true. In the case of 'one to one correspondence', it is the need to have a measure of size that applies to infinite sets. In the case of 'sibling', it is because sharing a parent is important for understanding the dynamics of social groups. In the case of phenomenal indistinguishability, it is to preserve transitivity. Our survey of those familiar issues will be one with an eye open on which properties it is good to have names for.

1.4 The Debate Over Knowledge

Imagine that you are at a workshop on knowledge and one speaker defends the view that knowledge is justified true belief (perhaps handling Gettier cases by urging that they aren't cases of justified belief). Another speaker defends the view that knowledge is reliably acquired true belief. They each likely present their view as a conceptual analysis, but their focus is, we may suppose, on the nature of knowledge itself.

They are disagreeing, that's obvious; but what *exactly* are they disagreeing about? One tempting answer is that there is a single property, that of being knowledge, and one party holds that it is identical with justified true belief and the other party holds that it is identical with reliably acquired true belief. This is the wrong way to think of the disagreement. For *which* property is it that one party holds is justified true belief and the other party holds is reliably acquired true belief? Justified true belief? No, *neither* party thinks that justified true belief is reliably acquired true belief. Reliably acquired true belief? No, *neither* party thinks that reliably acquired true belief is justified true belief. Some property distinct from these two? No, they *agree* that knowledge is one or the other. What then do they disagree about? They disagree about the property we use 'knowledge' for, or maybe the property we should use 'knowledge' for when we subject our usage to a certain amount of rational scrutiny.[3] On this way of looking at matters, it is easy to see why responses to

[3] The defence in Ayer (1956) of his version of the justified true belief account of knowledge is of this kind.

descriptions of Gettier cases matter. Responses to descriptions of Gettier cases tell us about the cases that subjects are happy to use ‘knowledge’ for; the responses are, that is, products of experiments to see what cases subjects are prepared to use the word ‘knowledge’ for.

I chose a defender of justified true belief and a defender of reliably acquired true belief to make the point about the real nature of the ubiquitous disagreement over knowledge, but of course our point does not depend on that choice. Consider, for instance, a disagreement between a defender of justified true belief and a defender of knowledge as a fundamental property that resists any kind of analysis. What does the first think that the second has got wrong? Not the nature of justified true belief, obviously. They think that the defender of knowledge as unanalysable has gone wrong, but not to the extent of holding that justified true belief is unanalysable, but justified true belief *is* knowledge according to them. Likewise, the defender of the unanalysability view of knowledge does not think that the defender of justified true belief thinks (absurdly) that something which is unanalysable is justified true belief. The way to make good sense of their disagreement is to see it as being about the property we pick out using ‘knowledge’—near enough, allowing for any adjustments in the light of the rational scrutiny we talk about above.

Now we have to wonder about the significance of the famously intractable ongoing debate over the nature of knowledge. If it comes down, in one way or another, to a question about word usage, it is of interest; but surely the focus of philosophers should be more on which properties are the important ones for epistemology. Questions about words would then be questions about which words are best to employ for the properties we discern as of most importance. This response we might well tag the conceptual engineering response. From this perspective, the real interest of Gettier cases is not so much that we—or a great many of us—are so reluctant to use ‘knowledge’ of them, while being happy to describe them as cases of justified true belief. The real interest is that they reveal something that it is easy to overlook: being a justified true belief is consistent with being a belief that falls a long way short of the gold standard, for it is consistent with getting something right by accident.[4]

In fairness to the famous ongoing debate about knowledge, some players in it have been clear that any analysis—as they typically phrase matters—of knowledge should allow us to see why knowledge matters. All the same, from the conceptual engineering perspective—as we are understanding conceptual engineering here, for a sense of the variety on offer, see Isaac (2021)—it is hard to justify the degree of concentration in the literature on finding a formulation that survives ingeniously constructed counter-examples, as opposed to finding the range of properties of importance to epistemology, properties which we can then tag in one way or another without worrying too much about the relationship between the tags and the word ‘knowledge’ as it used in English. It is also hard to justify an inference sometimes made from the failure to find a counter-example immune analysis of knowledge,

[4] For more on this theme, see Jackson (2011).

namely that there is a fundamental unanalysable property that is *sui generis* in the sense that it can only be captured in terms of the word 'knowledge' (or its cognates) itself. Why should a history of failure justify us in believing in an unanalysable fundamental property of especial importance to epistemology?

1.5 Swampman

Let Swampman be a molecule by molecule duplicate of me created by a massive, sudden aggregation of molecules that happened a moment or two ago. Swampman and I will be indistinguishable both in appearance and in regard to current and future interactions with our surroundings.

I have beliefs and desires. Does Swampman? Many are strongly of the opinion that he does, but there is substantial support for the view that he doesn't.[5] The second party urge that his states ('his' states, as they may say it, as 'he' may not be a person in their view) do not have the right kind of causal history to be about anything, where what is needed to have the right kind of causal history varies from author to author. It follows that Swampman's states lack content and, therefore, that none of his states are beliefs and desires.

For the conceptual engineer, it is a mistake to think that there is a major difference of opinion here about how I and Swampman are. What is true, and something everyone should agree on, is that I and Swampman are alike in an important way and different in an important way. As both ways are important, they deserve names. As it happens, I think that the respects in which Swampman and I are alike already have names; we do not need to coin new ones. In the case of English, the names are 'belief' and 'desire' etc. I also think that the agreed ways in which we differ have names. These names are to be found in the description of the case given near the beginning of this section, where the difference between my history and Swampman's is described.

Here is a way to think about the debate. Some geometry texts use 'ellipse' for closed plane figures that contain two internal points such that, for every point on their perimeter, the distance to one point plus the distance to the other is a constant. Other texts use 'ellipse' for closed plane figures that either contain two internal points such that, for every point on their perimeter, the distance to one point plus the distance to the other is a constant, or else contain one internal point such that, for every point on their perimeter, the distance to that point is a constant. Only the second count circles as special cases of ellipses. Surely, we conceptual engineers say, the wrong way to think about the difference is that the texts are disagreeing about the property of being an ellipse: *there is no property about which they disagree.* They are both clear that there are two different properties, and about the nature of the two properties. The difference is over which property they choose to give the

[5] Notably from Davidson (1987), to whom we owe the example.

name 'ellipse' to.[6] *Mutatis mutandis* for Swampman. There is, in this case also, no difference over the properties in play. Ergo, the difference is over naming.

Those who insist that Swampman lacks beliefs and desires will disagree with my view that we already have words for the respect in which Swampman and I are alike. (They will though agree about some of the right words for the respect in which we differ; they will have used words akin to them in their own articles about the historical difference between Swampman and me.) But this is a disagreement about words, not about the nature of the way in which we are alike. Indeed, I think this should have been obvious from the beginning. The Swampman case is easy to understand—that is, we knew from the beginning what is being discussed—and it was always obvious that carrying out experiments to settle whether or not Swampman had beliefs and desires would be a mistake in principle. Indeed, it is arguable that this 'it is really about words, not the metaphysics of Swampman' attitude fits well with how some who feel strongly that Swampman lacks beliefs and desires develop their case: they develop it by drawing on teleo*semantics*. I am not saying that they would agree with this way of characterising their view, but I am saying that it is suggested by some of their wordings.[7]

1.6 Narrow Content

The conceptual engineering question about narrow content is whether or not there is *a* narrow content property of belief and desire that is worth naming. Maybe our folk concept of content for belief implies that a belief's having such and such a content is a wide property of the belief, maybe it doesn't. Likewise for desire. But our question, as conceptual engineers, is whether or not there *exist* narrow content properties for belief and desire worth naming, or, to say it in the language of concepts, whether or not we need narrow concepts of content for belief and desire. My answer is that there are narrow content properties and that we do need narrow content concepts. I start with an explanation of what I mean—and what ought to be meant, it seems to me—by 'narrow' in this context.

Is being a good sunscreen a narrow or a wide property? In one sense, it is a wide property. It concerns how any substance that has the property is related to the Sun. And anyone who thought you could work out whether or not a substance was a good sunscreen by investigations restricted to the nature of the substance itself would be confused. In another sense, it is a narrow property: it is preserved under chemical duplication. This is why it is rational to buy generic brands instead of proprietary ones.

---

[6] Of course, conceptual engineers might want to express a view about which property is more important and so is best to use 'ellipse' for; but, unlike the prime number example we opened with, there is, as far as I know, no reason to favour one over the other.

[7] See, e.g., Papineau (2001).

It is clear that any property of belief and desire that might be a content property worth naming will be wide in the first sense. This is because *what* we believe and desire—that is, the contents of our beliefs and desires—is important for how we interact with our *surroundings*. But what about the second sense? Is there an important notion of content for which it is true that I and, as we might call him, generic I are alike in what we believe and desire?

The answer to this question is yes. Attributions of contents to beliefs and desires play an important role in predicting and explaining how people interact with their surroundings, as we in effect note above. We explain reaching for red wine rather than white wine, taking an umbrella, running for elected office, ... in terms of what's believed and desired. Of course, the preferred metaphysics of explanations in terms of what is believed and desired vary from theorist to theorist. Some talk in instrumentalist terms, others are robustly realist and posit sentences in a language of thought, others talk of maps in the head, and so on. But what is common ground is that what subjects believe and desire—the contents of their beliefs and desires—are important for explaining and predicting their behaviour. This means that there is *a* notion of content for which it is true that having content is a narrow property. For I and generic I behave in exactly the same way in all actual and possible circumstances. This means we do not differ in the content properties that are right for predicting and explaining behaviour in a set of *actual and possible circumstances*. If we did, there would be, somewhere in the set, a circumstance in which we do or would behave differently, and there isn't. Here's a way to put the key point. It would be good to have content properties that predict and explain my behaviour in each member of a set of circumstances, *independently* of whether or not the member is actual. But, as I and generic I behave alike in every actual and possible circumstance, these content properties will be shared between us; they will, that is, be narrow properties. And with our conceptual engineering hats on, we will want to give these narrow properties a name independently of whether or not they already have names. That's the sense in which conceptual engineering should make us supporters of narrow content.

I should distance what I have just been saying from an issue that sometimes comes up in discussions of Twin Earth. Suppose that I live on Earth and generic I lives on Twin Earth.[8] Many hold that our beliefs and desires differ. Mine are about H_2O, generic I's are about XYZ. These theorists also hold that our behaviour differs. For example, although we both on occasion use the same word 'water' to describe what we are reaching for, I am reaching for H_2O and generic I is reaching for XYZ. Others object that, in the relevant respect, our behaviour is the same—we both reach for the watery stuff. This issue is irrelevant to our key point above. What matters is the following. Suppose that we both go to bed on Monday night and, while we are unconscious, I am taken to Twin Earth and inserted into his bed, and generic I is taken to Earth and inserted into my bed. How I will behave on Twin

[8] I presume some familiarity with the fallout from Twin Earth, the example Putnam (1975) told us about.

Earth when I wake up on Tuesday is exactly the same as how generic I would have behaved had he remained on Twin Earth. And how generic I will behave on Earth when he wakes up on Tuesday is exactly the same as how I would have behaved had I remained on Earth. This is true independently of the issue about whether reaching for H_2O is or is not relevantly different behaviour from reaching for XYZ. Indeed, the debate over whether or not I on Earth and generic I on Twin Earth differ in what we believe when we express our beliefs using 'water' is by the way here. Maybe we do, maybe we don't. If we do, this tells us that there are *some* content properties that differentiate me from generic me, but our question as conceptual engineers is whether or not there also exist narrow content properties right for predicting and explaining behaviour in actual and possible circumstances, and the Twin Earth example in fact highlights the fact that there are. For I and generic I are exactly alike in our behaviour in actual and possible circumstances.

Narrow content raises a second question for conceptual engineers. How should we think of the truth conditions of beliefs with narrow content? Suppose that I have a belief with narrow content about how things are in front of me. Generic I will have the same (narrow) belief. How should we describe the belief that we share? The obvious answer is that the view I have about how things are in front of me is one and the same as the view he has about how things are in front of him—that is the sense in which we believe alike. But this implies that there is *a* sense in which our beliefs have different truth conditions. My belief's truth or falsity depends on how things are in front of me; his on how things are in front of him. But—surely—believing alike requires sharing truth conditions in *some* sense. There must be a sense in which our beliefs have the same truth conditions, and there is. My belief is true just if I belong to the class of subjects where things are a certain way in front of them; his belief is true just if he belongs to the class of subjects where things are that very way in front of them. What we share is the class to which we need to belong to in order for our beliefs to be true.[9] *Mutatis mutandis* for our desires: the sense in which our desires have the same satisfaction conditions is that the class to which we need to belong to in order for our desires to be satisfied is one and the same.

1.7 Proper Names

Why is it good to know the meaning of words like 'steep', 'road', and 'radioactive'? The reason is clear. It enables one to know how those who produce sentences like 'The road ahead is steep' or 'It is radioactive around here' take things to be. To the extent one judges them to be reliable, one can then respond appropriately. The precise connection between one's knowledge of the meanings of words and being able

[9] In my view, this is the right way to read David Lewis's talk of self-ascription of properties in Lewis (1979); see, e.g., Jackson and Stoljar (2020). Lewis's paper is an exercise in traditional conceptual analysis, but I am all but certain that he would have been sympathetic to conceptual engineering.

to recover how one's informants take things to be from the sentences they produce containing those words are matters of debate, but I insist (along with many) that the key to understanding why it is good to know what words mean is that one can recover how those who use the words take thing to be.[10] The problem one faces when in countries whose language one does not understand is precisely that one cannot recover from the words locals produce, where, for instance, they believe the nearest taxi stand is or which is the quickest way back to one's hotel.

These remarks tell us that we conceptual engineers can be happy with a notion of meaning along the lines just sketched. It makes it transparent why it would be good to know the meaning of 'radioactive' etc., and thereby why it would be good to have a word for it, namely, 'meaning'. And this is true independently of the extent to which the account is faithful to what is meant by 'meaning' in one or another extant account of meaning. Can we do something similar for proper names, can we, that is, give an account of their meaning that explains why it would be useful to know their meaning *on that account*? The answer to this question is yes, as we will now see.

Getting information about the world we occupy presents all sorts of challenges. Among them is the problem posed by there being so many items we need to keep track of: streets, rooms in office buildings, cities, people, countries, etc. We handle this problem in two ways. One is by having agreed ways of giving these objects distinctive tags: street names, numbers on houses and office doors, proper names for people and countries, etc. The agreed ways vary case by case: sometimes the tags are literally attached, as happens with numbers on office doors; sometimes the items are assigned through a process authorised by the state that typically leads to the item which gets the tag owning other items that have the tag physically—think of humans with their credit cards and birth certificates; sometimes there is an informal agreement to respond to a question like 'What is your name?' by producing the tag; and so on. We are reminded of the importance of these tags when a room number is left off a seminar announcement, and of the value of their being distinctive when we reflect on the advantages of having a middle name, especially when one's first and last names are common ones. Once these tags are in place, we can use them to pass on information about the items with the tags, as happens when we use room numbers to tell people where a seminar will be held or use a name to inform someone about who won the US Open.

There is, however, a problem with the tags: they are not *that* distinctive. A lot of people have the same name, there are many cities called 'London', etc. We get around this problem—and this is the second way heralded above—by exploiting our capacity to form beliefs about individual items relative to ourselves. This means that we are able to use *tokens* of tags, not just tag types, to collect information about particular items. Suppose that, back in Australia, I read the sentence 'London was hot yesterday' in the morning paper. Coming across this sentence allows me to collect information about a single object, because (1) although a lot of cities are tagged 'London', very many are not, (2) I am able to form a belief about the particular

---

[10] A view that goes back to Locke (1690: Book III, Ch. II, § 2) at least.

token in front of me, and (3) I know that this token is at the end of an information preserving chain that goes back to a single object tagged 'London'.

There is much more that could be said about this account of how names deliver information about particular objects. For example, what about historical figures who were tagged with names that are different from those we find in books today? How do the changed tokens in the books we read today deliver information about these people? All the same, I urge that the basic picture is a very attractive one and is in fact common knowledge—the folk know that something like this explains why names are so useful. And the basic picture allows us to answer the engineering question about names. For it is valuable to know that (when all goes well) tokens of names are at the end of information preserving chains that go back to an object tagged by the name or by some ancestor of the name.

How does this view relate to the direct reference account of the meaning of a name (Mill, 1867), which identifies the meaning of a name with its referent? Some defenders of that view will object that what I say above is an account of how a name like 'London' *gets* to have the meaning (= the reference, on their view) that it does. Semantics is being confused with meta-semantics. As a conceptual engineer, my response is to decline to quarrel over the semantics–meta-semantics distinction. Furthermore, echoing what was said earlier about the meaning of 'radioactive', I say, with my conceptual engineering cap on, that what is important is the way the account explains the utility of names, independently of whether or not it is the right account of the meaning of names on one or another conception of what meaning is. What is important is what explains why names are so valuable, and that is what the account outlined above addresses.[11]

1.8 The Role of Traditional Conceptual Analysis

There is, obviously, a sense in which being a supporter of conceptual engineering in the sense at play in this essay involves discounting the importance of conceptual analysis as traditionally conceived. The focus moves from—to say it in the language of concepts—understanding extant concepts to asking after the worth of those extant concepts and being open to finding new concepts or modifications of old ones that serve worthwhile purposes, or—to say it in the language of properties and words—the focus moves from understanding which properties are named by extant words to whether or not these are the right properties to name for the purposes at hand and being open to finding new properties to name using names that may or may not be old ones recruited to new tasks.

All the same, there are two important roles for traditional conceptual analysis that survive this change of focus. First, we know that the ways we classify the items

[11] The account of the value of names sketched here and the implications for the meanings of names have obvious connections with the causal descriptivism to be found in Kroon (1987) and Jackson (2010). The precise connection to Kripke (1980) is more controversial.

to be found in our world using *extant* words and concepts often capture properties of importance. Take, for instance, personal identity. We, and here I mean we folk, often attach a great deal of importance to whether or not a person identified in one context is the same person as a person identified in a different context. For example, we give special weight to the testimony of eyewitnesses to events—accidents, lectures etc. We do this because we know that testimony about an event from someone who was at the event—that is, from someone identical to someone at the event—has special value. And if someone does an especially good job of fixing our car, treating an illness or providing comments on a paper, we likely ask that person to fix our car, treat our illness or comment on a paper the next time the need arises. And so on. It is very hard to believe that practices like these—and I could have given many more examples—are irrational. We know, that is to say, that our everyday thought and talk about whether or not a person we interact with at one time is the same person as a person we interact with at another time must be picking out a property that matters. It *is* rational to ask for comments on a paper from someone who did a good job last time around. The testimony of eyewitnesses *is* valuable. But what property might it be?

We can think of traditional conceptual analysis as seeking to identify that property. The reflections on possible cases that dominate so much of the literature on personal identity are attempts to articulate this property. And because this property is important, this is obviously a worthwhile thing to do. What conceptual engineers, as we are conceiving of them, bring to the table is the following idea: of course, our folk thought and talk picks out an important property, but maybe there are a number of properties in the vicinity, as we might put it.[12] One concerns psychological continuities and dependences between person stages. That explains why two stages of the same person are likely to be similar in the quality of advice on a paper, how good they are at fixing one's car and so on, and why a later stage of a person is likely to have useful memories of events involving earlier stages of that person. A second property concerns the underlying material continuities that explain the first property. What then should we say in imagined cases where the two come apart, cases familiar from the personal identity literature? In these cases, the material that makes up an existing person is completely replaced but in a way that preserves all psychological continuities. Some insist that personal identity is preserved. Others object that this is to confuse identity with duplication. The conceptual engineer says that this is a question about our extant concept, the folk concept. It is an interesting question. There may be a determinate answer to it, or the matter may be indeterminate. But, once we see it for what it is, we can see that it is not the most important question. The most important question is which properties do the needed job, and we know that there are two properties that go together the way things actually are that do the needed job in the following sense: one continuity underpins the other. How

[12] See, e.g., Olson (2003) for a discussion of the possibilities here; what I say in the text that follows is a harmless (for the purposes at hand) simplification of a complex debate.

to label the imagined cases where they come apart is then a question of how best to distribute names for properties.

The second important role for conceptual analysis as traditionally conceived comes from the widely agreed fact that we have to start from somewhere. Engineering concepts is a good idea, but we need concepts to engineer to start with. Or, to say it in terms of properties, revising which properties are important requires that we have views about which properties there likely are to begin with. We can think of traditional conceptual analysis as directed at articulating the properties we acknowledge in how we take things to be in thought and talk at some given stage—the stage from which we can then proceed to launch attempts to do things better, or, in some cases, we may decide that our extant tools are just fine for the job at hand (change is not always called for).

1.9 Coda

I have been working with an understanding of conceptual engineering prompted by the examples given at the beginning. One might, however, think of conceptual engineering as engineering in the sense in which cars are the products of engineering or the sense in which revolutions get engineered. For we might allow ourselves to be changed in various ways so as to give us new concepts, perhaps with brain surgery or the implanting of an extra cone in our eyes that allows us to see more hues or …, where—to say it without using the term 'concept'—these changes would give us the ability to think, and maybe talk about, new properties. And of course, when we attend sets of lectures, conceptual engineering often happens in the sense that the lectures may *cause* us to believe in new properties and to acquire words for these properties. This more robust, as we might put it, sense of conceptual engineering does not, however, affect the basic message of this essay. Would conceptual engineering in this more robust sense be a good or a bad thing? The answer (setting aside the dangers of, e.g., brain surgery) will depend on the importance of the new properties and of our being able to think about them and name them—that is to say, on exactly the factors at play in this essay—and the uncontentious answer will be that sometimes it is a good thing. I say uncontentious, because we should all agree that it is desirable to identify, think about, and find names for the properties that are important for this or that endeavour. The contentious part of this essay is, of course, the role I give to this bit of common ground in my discussion of the debates over knowledge, Swampman, narrow content, and proper names.

Acknowledgments I am indebted to too many to name. Everything I say here has been influenced by discussions with colleagues and friends over many years.

References

Ayer, A. J. (1956). *The problem of knowledge*. Macmillan.
Davidson, D. (1987). Knowing one's own mind. *Proceedings and Addresses of the American Philosophical Association, 60*(3), 441–458.
Deutsch, M. (2020). Speaker's reference, stipulation, and a dilemma for conceptual engineers. *Philosophical Studies, 177*(12), 3935–3957.
Goodman, N. (1977). *The structure of appearance*. Reidel.
Isaac, M. G. (2021). What should conceptual engineering be all about? *Philosophia, 49*, 2053–2065. https://doi.org/10.1007/s11406-021-00367-x
Jackson, F. (2010). *Language, names, and information*. Wiley-Blackwell.
Jackson, F. (2011). On Gettier holdouts. *Mind and Language, 26*(4), 468–481.
Jackson, F., & Stoljar, D. (2020). Understanding self-ascription. *Mind and Language, 35*(2), 141–155.
Kripke, S. (1980). *Naming and necessity*. Blackwell.
Kroon, F. (1987). Causal descriptivism. *Australasian Journal of Philosophy, 65*(1), 1–17.
Lewis, D. (1979). Attitudes de dicto and de se. *Philosophical Review, 88*(4), 513–543.
Locke, J. (1690). *An essay concerning human understanding*. Thomas Basset.
Mill, J. S. (1867). *A system of logic*. Longmans.
Olson, E. T. (2003). Personal identity. In S. Stich & T. Warfield (Eds.), *The Blackwell guide to the philosophy of mind* (pp. 352–368). Wiley-Blackwell.
Papineau, D. (2001). The status of teleosemantics, or how to stop worrying about Swampman. *Australasian Journal of Philosophy, 79*(2), 279–289.
Putnam, H. (1975). The meaning of 'meaning'. In *Mind, language and reality* (pp. 215–271). Cambridge University Press.

Chapter 2
Conceptual Engineering, the Value of Knowledge, and the Value of Understanding

Jennifer Nado

Epistemology, for a large chunk of the twentieth century, was dominated by attempts to produce a conceptual analysis of knowledge. As everyone knows, it didn't go too well. Gettier's (1963) introduction of his eponymous counterexamples catalysed a flurry of attempts to pin down the elusive 'missing ingredient' that JTB apparently lacked. Yet to this day, despite reams of paper and gallons of ink spent, no solution has been found. Within the last quarter-century or so, there have been growing signs that the discipline has finally started to grow weary of the chase. Linda Zagzebski (1994), for instance, has argued that the Gettier problem is simply an inescapable consequence of fallibilism—that it cannot be 'solved'. Fans of 'knowledge first' approaches (following Williamson, 2000) have, meanwhile, suggested that knowledge may simply be basic and unanalysable. Perhaps this looming sense of futility explains why epistemology's primary focus has slowly started to shift away from the analysis of knowledge, to embrace a variety of more general epistemological topics such as epistemic value or the ethics of belief.

The knowledge-centred nature of epistemology may, in fact, be a fairly recent development. Ancient Greek philosophy gave pride of place to *epistêmê*—a term that is sometimes translated as 'knowledge', but arguably comes much closer to 'understanding'. Twentieth-century epistemology, by contrast, gave understanding hardly a second glance. Was this simply a mistake? Might *understanding* ultimately be a much more valuable, much more theoretically interesting locus for our epistemological inquiries? Within the past few years, interest in the epistemic state of understanding has undergone a resurgence, leading authors such as Linda Zagzebski (2001, 2009), Jonathan Kvanvig (2003, 2009), Catherine Elgin (2006), Duncan Pritchard (2009, 2014) and more to propose that, to one degree or another, understanding ought to take over some portion of knowledge's philosophical market share.

J. Nado (✉)
University of Hong Kong, Hong Kong, China

M. G. Isaac et al. (eds.), *New Perspectives on Conceptual Engineering - Volume 2*, Synthese Library 514, https://doi.org/10.1007/978-3-031-98524-9_2

The suggestion that epistemology should re-orient its focus towards understanding—a proposal we might call the 'Refocusing Thesis'—can be taken either in a weaker or in a stronger sense. In the weak sense, it simply amounts to the claim that the study of understanding has been unjustly neglected. This is compatible with knowledge retaining its current primary status within epistemology, though perhaps with a somewhat reduced monopoly on our attentions. A stronger version of the Refocusing Thesis might claim that understanding should largely displace knowledge—that it, rather than knowledge, should form the heart of epistemology as a discipline. And, of course, there's a spectrum of possible positions in-between.

Calls for one or another degree of Refocusing form one theme in the recent bumper crop of understanding literature; another involves exploration of a variety of questions regarding understanding's nature and relation to knowledge. What is the object of understanding? Is understanding merely a form of propositional knowledge? Does understanding require belief? Truth? The list goes on.[1] Typically, however, the methods of authors in the understanding literature more or less resemble the Gettierology-soaked debates of the later twentieth century. There's much bandying about of counterexamples, much attention paid to the contours of the concept as it is used by laypersons or non-philosopher academics such as scientists. In other words, a lot of the current understanding literature looks like (good?) old-fashioned conceptual analysis.

Which is odd, really. If any moderately robust version of the Refocusing Thesis is correct, our knowledge-concept never really deserved the theoretical attention that we lavished upon it—though the term 'know' is one of the most frequently-used words in the English language, the concept it reflects isn't really of particular philosophical interest. But if our most-used epistemic concept is really not all that it's cracked up to be, shouldn't we be at least a little sceptical about the theoretical importance of our folk concept of understanding as well? If epistemology needs to refocus its attentions, why limit our targets to the epistemic concepts in our current repertoire?

I'll suggest that we should instead consider the Refocusing Thesis from a *conceptual engineering* perspective. 'Conceptual engineering' refers to the practice of seeking to evaluate and improve our concepts, rather than merely to analyse them. It involves revision, rather than description—and it permits the introduction of entirely new concepts when our current conceptual inventory is found lacking. When applied to the Refocusing Thesis, a conceptual engineering approach would involve not only a comparison of the relative merits of the concept KNOWLEDGE[2] and the concept UNDERSTANDING, but also a consideration of whether either concept could be improved. Or, for those skeptical about the feasibility of

---

[1] I focus in this chapter on authors who have argued in favour of refocusing our philosophical attentions towards understanding, but see Hannon (2021) for a good recent overview of literature on some of the other issues just mentioned.

[2] I'll be using capital letters to denote concepts, as opposed to linguistic expressions or the phenomena to which they refer. KNOWLEDGE is a concept; 'knowledge' is the linguistic term that expresses that concept; knowledge is the epistemic state to which 'knowledge' refers.

modifying concepts, whether these concepts could be *replaced* by similar, 'nearby' concepts that would be even more worthy of our philosophical investigations. Sure, perhaps UNDERSTANDING is of more theoretical interest than KNOWLEDGE—but oughtn't we to consider whether slightly different concepts, KNOWLEDGE* or UNDERSTANDING*, might be more interesting still?

Current advocates of conceptual engineering are a diverse group, and there's much disagreement over the conditions for successfully engineering a concept (and over much else besides!). One of the more popular approaches—the one which I will adopt in this chapter—might be broadly termed 'practical', 'pragmatic' or 'functional'.[3] A practical/pragmatic/functional approach to conceptual engineering (henceforth, I'll call it the 'practical' approach) is a view that holds that the success conditions for conceptual engineering should in some way or other focus on the functions or roles that our concepts play in our cognitive lives.[4] Thus, conceptual improvement is framed in terms of an increased ability to effectively serve those roles, and thereby to help us attain the various goals and aims those roles support. As an example of the sort of 'roles' our concepts might play, consider the concept CONSCIOUSNESS: apart from its obvious role in explaining certain movements of various medium-sized objects in our environment, CONSCIOUSNESS is also appealed to in order to flag appropriate targets of moral concern, and to help us make sense of our own subjective experiences.

Applied to the Refocusing Thesis, the practical approach to conceptual engineering would suggest the following strategy. First, we consider the role or roles played by our current set of epistemic concepts. What are these concepts *used* for? How do they contribute to our ability to successfully navigate the world? Next, we consider how well our current concepts serve these roles. We then consider whether these concepts could be improved or replaced with 'successor' concepts that *better* serve the identified roles. Finally, we consider whether the practical utility (or lack thereof) of the various concepts we've considered justifies an increase (or decrease) in the theoretical focus said concepts are given. On this strategy, then, we're not merely asking whether our current concept UNDERSTANDING can play various needed epistemic roles better than our current concept KNOWLEDGE; we're also

---

[3] For prominent examples of this approach, see Prinzing (2018), Haslanger (2020), Simion and Kelp (2020), Thomasson (2020).

[4] Cappelen (2018) argues against the idea that concepts have functions. Much of his argument targets practical accounts which present concepts as having a 'core' or 'central' function, noting that 'the only universal, i.e., stable, function of a concept 'C' is to denote Cs.' (Cappelen, 2018, p. 183). However, the identification of a single function active in all contexts is not required for a coherent practical account; as we'll see, KNOWLEDGE likely serves several functions which are only active in particular contexts (such as, for instance, determining the permissibility of an assertion). Cappelen certainly accepts that 'people have goals and aims and purposes when they use words [and concepts] on particular occasions'; it's not clear to me that anything more substantive is needed than this. A practical account can be formulated in terms of the purposes for which we use our concepts, with the corresponding success conditions spelled out in terms of how well a concept helps us in those aims.

asking whether it might be possible to engineer up successor concepts KNOWLEDGE* or UNDERSTANDING* which can play those roles better still.

In the sections to come, I'll follow this general approach. I'll argue that, for many of the salient roles that epistemic concepts play in our lives, KNOWLEDGE seems to be more suited to effectively fill those roles than UNDERSTANDING. For instance, KNOWLEDGE seems better suited than UNDERSTANDING to pick out a 'stopping point' for inquiry under most normal circumstances, and it seems more suited as a way of identifying a minimal threshold for warranted assertion and action. But that's not to say that KNOWLEDGE is *perfectly* well-suited to these roles; for instance, as some fans of Refocusing note, it's not necessarily clear that KNOWLEDGE would serve the relevant purposes better than plain old JUSTIFIED TRUE BELIEF. However, from a conceptual engineering perspective, the crucial question is whether the ideal concept for filling these roles would be more KNOWLEDGE-*like*, or more UNDERSTANDING-like—that is, whether KNOWLEDGE or UNDERSTANDING would serve as a better 'jumping-off point' for revisionary efforts. And here, I'll suggest that it's very clear that KNOWLEDGE takes the crown—or, more accurately, passes it to KNOWLEDGE* rather than UNDERSTANDING*. In other words, the proper response to the alleged inadequacy of KNOWLEDGE is to *re-engineer it*, not to abandon it in favour of UNDERSTANDING.

However, I'll also claim that we can identify other, less common epistemic roles for which UNDERSTANDING seems likely to serve better than KNOWLEDGE. For instance, while we often attribute knowledge to 'flag' a person as a reliable source of testimony (as suggested by Craig, 1990), attributing understanding often serves to 'flag' a person as capable of resolving various problems related to the area she understands. But again, it's possible, even likely, that some nearby concept UNDERSTANDING* would serve yet better. Thus, I'll ultimately suggest that while KNOWLEDGE (or more likely, some suitable successor) ought to retain pride of place in our epistemological inquiries, UNDERSTANDING (or its successor) may well deserve a greater share of attention than it currently receives.[5] That is, while the strong version of the Refocusing Thesis is false, the weak version may well be true.

---

[5]Do epistemologists investigate KNOWLEDGE, or knowledge? I'm tempted to say 'both'. KNOWLEDGE refers to, or has as its content, or represents, knowledge. So we might characterize the philosophical enterprise as investigating knowledge, or as investigating what KNOWLEDGE refers to, or as investigating the content of KNOWLEDGE, or what have you. Tied in with this question is the question of whether conceptual engineering investigates concepts, or phenomena in the world—a question to which I'm inclined to give a similar answer. The aim of a conceptual engineering project (on the practical approach) is to find a way of 'carving up' reality which most effectively serves a target role or roles. So we're looking for a suitable phenomenon in the world—that is, a category, or a 'piece' of reality—but we're also looking for a tool for representing that phenomenon—that is, a concept.

2.1 The Case for Refocusing

Before embarking on our engineering-based consideration of Refocusing, let's first pin down the Refocusing Thesis a little more clearly, and then briefly consider the arguments for Refocusing that more 'traditional' fans of understanding have presented. The authors we'll consider here are often fairly coy about the degree of Refocusing they take their arguments to support; and so, rather than aim to pin down the exact positions of individual authors, I suggest we rather abstractly consider the degree of Refocusing that arguments along these lines might motivate. We'll then examine whether a conceptual engineering approach affects our assessment.

Let's start with the Refocusing Thesis itself. I've glossed this as the claim that 'epistemology should re-orient its focus towards understanding', and I've made passing reference to 'theoretical importance' and 'theoretical interest', and the amount of 'theoretical attention' that an epistemic concept may or may not deserve. Essentially, I take the Refocusing Thesis to be a normative claim about the division of cognitive labour within the discipline of philosophy. To advocate for Refocusing is to suggest that there ought to be more articles and books about understanding, and fewer about knowledge; that we philosophers should allocate more cognitive effort towards exploring the characteristics of understanding and the roles it serves in our epistemic lives; and so forth. For a traditionalist, this might amount to a suggestion that we spend more time and effort on the conceptual analysis of UNDERSTANDING. For a fan of both Refocusing *and* conceptual engineering, the suggestion might be that we spend more time evaluating UNDERSTANDING's suitability for the various epistemic roles that we take to need filling, and subsequently tinkering with its features to improve its performance. As I'm using the terms, a concept's theoretical importance determines how much theoretical attention, measured in terms of share of cognitive labor, it deserves.[6] Just what it is for a concept to have theoretical importance will be a key issue in what follows; we'll return to it shortly.

Let's turn now to arguments for the Refocusing Thesis. Here the dominant argument, and the one with perhaps most interest from a conceptual engineering perspective, is the claim that knowledge simply does not exhibit the sort of unique *value* that we'd expect from the centrepiece of epistemological inquiry. The basic puzzle here extends all the way back to Plato's *Meno*. In that dialogue, Plato famously notes that knowing the correct path to Larissa provides no practical benefit over merely having a *true opinion* about the correct path to Larissa; in either case, one will get to one's destination. Knowledge indeed has practical value, then, but not obviously more so than true belief—and the latter would presumably be at least somewhat more amenable to conceptual analysis. So why torture ourselves over the profoundly elusive analysis of knowledge?

[6] I'll also sometimes use 'theoretical interest' in roughly the sense of 'deserving of theoretical attention'.

Duncan Pritchard (2007) has labelled the basic problem of explaining knowledge's superior value over true belief the 'primary value problem'; he distinguishes this from the 'secondary value problem' of explaining the particular value that knowledge adds to *justified* true belief, or to any other subset of its proper parts. Pritchard explicitly suggests that knowledge's pride of place within contemporary epistemology is premised on an assumption that it is possible to solve these value problems: "It is a widespread pre-theoretical intuition that knowledge is distinctively valuable. If this were not so, then it would be simply mysterious why knowledge has been the focus of so much of epistemological theorizing" (Pritchard, 2009, p. 19). If this is right, then absent some good reason to think that suitable answers are forthcoming, the post-Gettier preoccupation with the analysis of knowledge starts to look like a major misstep.

And indeed, identifying a solution to Pritchard's value problems has proved harder than we might initially expect. Jonathan Kvanvig, who has argued extensively against the presumption that knowledge has unique value, lays out the options as a dilemma (Kvanvig, 2003). Knowledge's value might be due to extrinsic factors—presumably, the practical utility it provides in achieving our goals. Or, its value might be intrinsic to it. On the first option, however, it's difficult to see how knowledge provides any value beyond that of mere true belief. Plato's own solution to the *Meno* puzzle was that knowledge is more useful because it is 'tethered'—more stable, and less likely to evaporate under doubts or receipt of new evidence.[7] But, as Kvanvig points out, knowledge does often disappear, whether due to simple forgetting, lack of confidence, or the presence of misleading evidence. Our most stubbornly stable beliefs, by contrast, may often be our least rationally held ones—'dogmatically-held true belief', it seems, would serve us better in the stability department.[8]

On the other hand, we might attempt to identify knowledge's value as something intrinsic to it. We do, Kvanving claims, seem to intrinsically value true belief—even independently of its practical utility. But it's not clear what *knowing* adds to this value. Kvanving singles out reliabilist accounts as facing particular difficulty here, since it would seem that any value granted by the reliability of knowledge is 'swamped' by the value already provided by truth. Kvanvig suggests that a subjective, internalist account of justification might hold more promise, in that it might appeal to value arising from the relative transparency of such justification to reflection. But even were this to succeed, justified true belief is of course not equivalent to knowledge—and there's simply no promising account of the added value provided by Gettier-proofing. At best, then, justified true belief has unique value, but knowledge adds nothing more. Kvanvig's ultimate suggestion is that understanding, rather than knowledge, has unique value. Understanding involves the sort of subjective justification that Kvanvig holds to be valuable, but unlike knowledge (Kvanvig

[7] For a contemporary defense of this strategy, see Williamson (2000).

[8] Kvanvig has further intriguing arguments here, but the basic arguments just discussed are enough for current purposes; we'll return to his discussion of knowledge's role in regulating assertion and action later in the chapter.

claims), understanding requires no 'extra' anti-luck condition. There is, then, no equivalent to Pritchard's 'secondary' value problem for understanding.

Pritchard (2008), meanwhile, also provides arguments indicating that an account of knowledge's distinctive value is not likely to be forthcoming. The most promising approach to explaining knowledge's value, Pritchard claims, appeals to its status as a cognitive achievement; cognitive achievements, after all, intuitively possess non-instrumental, intrinsic value. But unfortunately, not all knowledge reflects substantive achievement. As Jennifer Lackey (2007) has pointed out, knowledge is often obtained through fairly unreflective acceptance of testimony. For instance, I may come to know how to get to Sears Tower by asking a random passer-by. It's hard to identify any particularly credit-worthy intellectual achievement inherent in that sort of knowledge-acquisition. On the other side of the coin, certain Gettier cases exhibit cognitive achievement without knowledge—fake barn cases being the prototypical example. So cognitive achievement plausibly has intrinsic value, Pritchard claims, but knowledge is not equivalent to cognitive achievement. *Understanding*, by contrast, does seem to constitutively involve cognitive achievement. If that's right, then it is understanding, not knowledge, that possesses unique value.

There are other routes to motivating the claim that understanding possesses a unique value that knowledge lacks, but they needn't detain us here—I'm inclined to accept Kvanvig and Pritchard's versions, at least for the sake of argument. Let's suppose, then, that knowledge simply may not exhibit any particular unique value that exceeds that provided by justified true belief. Understanding, meanwhile, plausibly holds unique intrinsic value. And what's more, it plausibly provides greater value than knowledge does; given the choice between knowing something and understanding it, most of us would opt to understand. Understanding, as many authors note, involves what seems to be a deeper, more comprehensive grasp of reality. It involves an "appreciation or grasp of order, pattern, and how things 'hang together'" (Riggs, 2003, p. 217); a "grasping of explanatory and other coherence-making relationships in a large and comprehensive body of information" (Kvanvig, 2003, p. 192).[9] It is, in other words, arguably a 'higher' epistemic state than knowing.

What degree of Refocusing do such considerations support, if any? If we take the presence of unique value to be our primary guide to theoretical importance, as Pritchard's argument suggests, then *prima facie* the case for keeping knowledge as our primary focus starts to look fairly shaky—making room, presumably, for understanding. Nonetheless, it's not at all clear to me that the heightened value understanding exhibits translates directly to theoretical importance. Nor is it clear to me that knowledge's lack of *unique* value makes it less worthy of epistemological attention than understanding. One trouble here is that the 'value' being invoked here is a bit of a slippery notion. The sort of 'extrinsic' practical value Kvanvig refers to is fairly clear—and it fits fairly nicely with the practical conceptual engineer's notion

[9] See also Bourget (2017) for discussion of the mysterious 'grasping' that often features in characterizations of the difference between knowledge and understanding.

that good concepts help us attain our goals. The 'intrinsic' value of knowledge, on the other hand, seems largely a matter of subjective intuition. But perhaps that's fine; after all, we don't feel the need to justify the fact that we value pleasure, happiness, love, and so forth.

A second complication, however, arises due to the fact that the value of a *concept* need not be equivalent to the value of the *phenomenon* that concept picks out. As an obvious example, we don't very much value depression (quite the opposite!), but the concept DEPRESSION has great practical (that is, extrinsic) value insofar as it facilitates identification and subsequent treatment of a common health issue. Even when the relevant phenomenon *is* one that we value, its value might be either lower or higher than the value of its corresponding concept. That is to say, the most valuable concepts are not necessarily concepts of the most valuable things. As an example: most of us would prefer to be given a bottle of champagne than a bottle of water, but if I had to choose between possessing the *concept* CHAMPAGNE or the *concept* WATER, I'd certainly pick the latter. This holds just as well for the case of knowledge and understanding. It's true that most of us would prefer to understand than to merely know; but if understanding is fairly rare, while knowledge is fairly common, then the ability to accurately pick out knowers might be more (extrinsically, practically) valuable than the ability to accurately pick out understanders. And thus KNOWLEDGE might be more valuable than UNDERSTANDING even while understanding is more valuable than knowledge.

What's more, the theoretical importance of a concept does not necessarily depend on its referent's displaying *unique* value. This is particularly true when we adopt a conceptual engineering approach. Many arguments against knowledge's unique value point out that knowledge seems to have no obvious advantage over justified true belief. But if that's right, then all it seems to show is that KNOWLEDGE possesses an unneeded feature—the conceptual equivalent of a vestigial appendix, perhaps. This might motivate a revisionary version of KNOWLEDGE (call it KNOWLEDGE*) that simply excises the impotent extra condition—a return, in other words, to a vintage JTB account. This wouldn't remove the need for additional epistemological work on KNOWLEDGE*; characterizing the nature and extent of the required justification, for instance, is far from a trivial matter. And there's still a clear sense in which we could say that KNOWLEDGE is a useful concept, even if it is ultimately to be superseded by a more streamlined successor: KNOWLEDGE, in the case just imagined, would be serving as the model or jumping-off point for our eventual revisionary efforts. Indeed, that's just what I plan to argue in the sections below—KNOWLEDGE deserves to remain central to the discipline of epistemology, if only as a model for a successor that better serves the primary roles for which our epistemic concepts are used.[10]

How does all this translate to degree of deserved theoretical attention? I'd suggest that we ought to distribute our theoretical attentions largely in proportion to the

[10] I should note that I do not intend here to endorse any particular 'similarity' requirement on conceptual engineering. A concept may serve as a model or inspiration for a successor that diverges from it substantially in extension or in intension.

value of the relevant concepts, rather than their objects. This is partly just a reflection on actual practice—the converse strategy fails to make sense of why psychiatry studies depression, or why chemists spend more time on water than on champagne. From the perspective of a practical conceptual engineer, however, a focus on the value of the concept (rather than its object) also simply jives well with the overall characterization we've given to conceptual engineering success. The whole point of the endeavour, at least on the practical view, is to engineer up useful *concepts*—that is, ones that are extrinsically valuable due to the roles they serve and their contribution to our various purposes and aims. This concept-driven notion of theoretical importance is, I should add, compatible with claiming that the ultimate object of philosophical study is the *phenomenon* itself rather than its concept. We might say, for instance, that it is important for scientists to work out the essential features of depression, because doing so will result in a well-defined concept DEPRESSION that will have extrinsic value for diagnosis and treatment.

Applying all of this to the epistemic case, I propose that the degree of theoretical attention we ought to distribute to understanding/UNDERSTANDING and to knowledge/KNOWLEDGE will be (again, largely)[11] proportional to the relative values of UNDERSTANDING and of KNOWLEDGE. The values of these concepts may track the values of their objects, but they may not—and that goes double for successor concepts. We might, for instance, find that a bespoke concept KNOWLEDGE* with a lower justification threshold than KNOWLEDGE would be better suited to certain roles currently played by KNOWLEDGE—rather like how a job recruiter might find it useful to lower a job ad's experience requirement to offset a paucity of qualified applicants. In such a case, though knowledge* would be a 'lower' epistemic state than knowledge, and thus presumably less valuable, KNOWLEDGE* would arguably be more valuable than KNOWLEDGE.

2.2 The Roles Our Epistemic Concepts Play

The best way to assess the relative values of KNOWLEDGE and UNDERSTANDING, I'd suggest, is to examine the roles each concept plays in our cognitive lives. Let's begin with KNOWLEDGE. Why do we ascribe knowledge? What purpose, or purposes, are served by picking out certain mental states as instances of knowing? There are, in fact, a variety of plausible answers to these questions that fall fairly directly out of existing literature in epistemology.[12] In this

[11] I emphasize 'largely' here because other factors will certainly contribute to our assessment of the ideal distribution of philosophical cognitive labor, For instance, difficulty will play a role—it may be more worthwhile to spend time on 'low-hanging fruit', than to invest substantial cognitive effort into a topic where progress is hard-won. The exact tradeoffs here lie beyond the scope of this chapter, unfortunately, so I'll focus my efforts on value.

[12] My list here is directly inspired by a similar survey of knowledge's functions/purposes laid out by Fassio and McKenna (2015).

section, I'll survey several potential roles that KNOWLEDGE might play in our cognitive lives—several reasons, that is, why we might deem it useful to determine whether or not a mental state falls under that concept. It's a matter of controversy whether or not KNOWLEDGE in fact *does* play these roles; but ultimately, the primary goal of this section is to identify certain important purposes whose fulfilment calls for use of *some* epistemic concept. We can then proceed to inquire into the further question of whether KNOWLEDGE, UNDERSTANDING, or some custom-made successor concept would be the best tool for the job.

The first role or purpose that immediately suggests itself is the role of regulating permissible instances of assertion. There's a substantial literature directed at the question of whether knowledge is the 'norm of assertion'; on one common formulation, this is the question of whether one may assert p only if one *knows* that p.[13] It's certainly clear that we do have *some* norm of assertion in place, and it is at least *prima facie* plausible that this norm is formulated in terms of knowing—after all, we commonly object to epistemically flippant assertions with the phrase "how do you know?" Moreover, there is evident utility to employing *some* sort of epistemic concept to regulate assertion in this manner. Doing so guides a concept-user both in deciding what she may permissibly assert, as well as in evaluating and potentially criticizing the appropriateness of the assertions of others. Community-wide employment of a concept for this purpose puts into place a default presumption of epistemic quality that helps listeners determine how much credence to grant to testimony.

A second, closely related role that KNOWLEDGE appears to play is in flagging appropriate sources of information, a suggestion that originally comes from Craig (1990). A very large proportion of our beliefs about the world are acquired through testimony, and as such we have a need to distinguish good sources of testimony from bad. Knowledge-attribution plausibly serves to so distinguish: by attributing knowledge that p to a subject, I mark out said subject as a good source of testimony as to whether p. Again, we can see evidence for this usage in everyday language—consider the frequency with which we say things like "Ask Sarah, she'll know" or "Does anyone know what time we are meeting?". And again, though we might debate whether KNOWLEDGE really does play this role, what's important for current purposes is that there is practical utility in having *some* epistemic concept play this role.

Third, more than one author has suggested that knowledge serves as a 'stopping point' for inquiry, in the following sense: when one knows that p, one may reasonably close inquiry as to whether p. This role has been defended by e.g. Kappel (2010), Kelp (2011), and Rysiew (2012). The conversational evidence for KNOWLEDGE's occupation of this role is a bit more subtle, but can be seen in the naturalness of claims like "I don't need to check the timetable, I already know when the train leaves", or "I don't know whether we have any milk, I should go look". Once again, it is fairly easy to see the usefulness of having a concept to mark this

[13] I've formulated the norm as a necessary condition, rather than a sufficient or necessary-and-sufficient one. However, see Brown (2010). Well-known defenders of a knowledge norm for assertion include e.g. Williamson (1996, 2000) and DeRose (2002).

sort of stopping-point. A concept serving this role might serve to guide our own 'epistemic actions'[14] (e.g., reflecting or seeking out additional evidence), as well as guiding blame or praise of others' epistemic actions.

Finally, KNOWLEDGE has also been held to have a role in regulating practical reasoning. Several authors hold that knowledge is the 'norm of action', in the sense that one should act on p only if one knows that p.[15] As with our previous examples, everyday use of 'know' provides some plausibility to this claim. Consider, for instance, statements like "you shouldn't climb that rickety ladder, you don't know whether it's safe". And, yet again, it's easy to see the practical utility in having a concept play this role; a concept playing such a role could be utilized to both guide our own practical reasoning, and to license praise and blame of the practical reasoning of others.

There are more potential roles we might identify for KNOWLEDGE, as well. For instance, Williamson (2000) claims that knowledge plays an ineliminable role in the prediction and explanation of behaviour. Machery (2020) points out that attribution of *mens rea*, both in a legal context and in informal attributions of moral blame, crucially invokes KNOWLEDGE. He also notes its role in religion—e.g. in concepts of divine omniscience—and points to the early emergence of the knowledge-concept in childhood development as indication of its central importance in our cognitive life. But for current purposes, the four roles we've examined above will serve well enough.

Let me re-emphasize—I don't mean to argue conclusively that KNOWLEDGE *does* serve these roles. That would require much more argumentation than I have given, and would require dealing with counter-arguments such as those of Kvanvig, who rejects knowledge's status as the norm of assertion and action, as well as the aim of inquiry, during his discussion of potential sources of practical, external value for knowledge.[16] What I *do* mean to claim is two-fold. First, I claim that KNOWLEDGE *prima facie appears* to serve these roles, as evidenced by instances of the usage of 'know' and its variant forms in ordinary conversational contexts.[17] Second, I claim that reflection indicates that *some* epistemic concept *ought* to serve these roles; that is, that there is high practical utility in having a concept (or concepts) that picks out an accepted threshold (or thresholds) for assertability, actability, inquiry-closure, and so on. What is ultimately of interest, from a conceptual engineering perspective, is what sort of concept *ought* to fill these roles. We'll return

[14] I use the term 'epistemic actions' here to indicate belief-influencing activities that are within our intentional control. Thus, while there's dispute about whether one can will oneself to believe, one can certainly will oneself to go looking for more evidence.

[15] See e.g. Fantl and McGrath (2002), Hawthorne (2004), Stanley (2005), and Hawthorne and Stanley (2008).

[16] Kvanvig claims that there's no reason to think that knowledge, rather than justified belief, is required for assertion and action; he also suggests that inquiry merely aims at truth.

[17] For a discussion of how well philosophers' claims about the ordinary usage of 'know' reflect data from linguistic corpora, see Hansen et al. (2021).

to that question shortly—first, we must also examine the potential roles played by UNDERSTANDING.

We've just discussed the role KNOWLEDGE appears to play in 'flagging' appropriate sources of testimony. UNDERSTANDING appears to play a somewhat similar, but more restricted role—it appears to serve to flag persons who are potential problem-solvers for the domain that they understand.

> When we want an expert about a problem, we consult a person who has understanding of the subject matter, since such a person is likely to be a reliable problem solver...what enables her to figure out the solution to the problem is her understanding of a complex chunk of the world, not simply her knowledge or true beliefs about that portion of the world. (Zagzebski, 2001, p. 245)

Similar sentiments, backed by empirical evidence from the judgments of non-philosophers, can be found in Wilkenfeld et al. (2016). In particular, the possessor of understanding seems to be better-situated than a mere knower to potentially resolve problems that are novel or unexpected. Suppose you are diagnosed with a rare and unusual cancer. Who would you prefer as your oncologist: a doctor who has merely memorized a number of textbook facts about cancer without understanding, or a doctor who deeply understands the mechanisms behind cancer growth and development?

A second role UNDERSTANDING may play is a bit more specialized—it seems to be frequently invoked in our characterization of the aims of science. Intuitively, science is in the business of *understanding* reality, not just gathering lists of factual knowledge. Understanding seems well-suited to facilitate prediction and explanation; which, of course, are very commonly supposed to be central aims of scientific inquiry. Indeed, it's extremely common to link understanding and explanation. The relationship between understanding and scientific explanation has been the target of a quite extensive literature since even before the recent re-emergence of interest in understanding within more traditional areas of epistemology.[18]

Understanding may also have explanation-related roles to play outside of the domain of science. Turri (2015), for instance, suggests that understanding is the 'norm of explanation'—in offering an explanation, one represents oneself as an understander. Hannon (2019), meanwhile, has put forth a Craig-inspired suggestion that the primary purpose for understanding-attribution is to flag testimonial sources of explanation. As with KNOWLEDGE, there's plenty of room to debate whether UNDERSTANDING in fact does play these roles, and moreover whether there might be some improved, UNDERSTANDING-like concept that could play these roles yet more effectively. A full investigation would take much more space than is available to us here; suffice it to say that there seems to be a *prima facie* good case to be made that UNDERSTANDING does currently serve certain unique roles that KNOWLEDGE does not.

[18] See e.g. Hempel (1965), Trout (2002), de Regt (2009), Khalifa (2012).

2.3 KNOWLEDGE Versus UNDERSTANDING as a Model for Concept Design

As the previous section indicates, reflecting on the various contexts in which we commonly ascribe knowledge and/or understanding can be a helpful first step in piecing together an idea of what we want our epistemic concepts to do; in this way, it can serve as a preliminary to a conceptual engineering project aimed at revamping our epistemic concepts. Said reflection can draw our attention to a variety of purposes for which epistemic concepts might be profitably used; we can then consider how to best tweak our conceptual practices to most effectively fill these roles. Let's move on to that task now.

In this section I'll first argue that, even if KNOWLEDGE is not *ideally* suited to play any of the roles we earlier found it to serve, UNDERSTANDING is quite clearly a *much* poorer match. I suspect that the best possible concept for these roles will be custom-tailored, at least to some degree—what I really mean to suggest, then, is that KNOWLEDGE seems much better suited than UNDERSTANDING to serve as a *model* for a custom-built assertion-regulating concept; for a custom-built inquiry stopper; and so forth. If this is right, then epistemologists will have good reason to spend a substantial amount of their theoretical energy on a concept that is at least 'KNOWLEDGE-like'. Consequently, shifting our focus from KNOWLEDGE to UNDERSTANDING would impair our engineering efforts, at least with regard to the sorts of purposes KNOWLEDGE currently serves.

However, on the other side of the coin, I'll go on to argue that KNOWLEDGE is arguably more poorly suited than UNDERSTANDING to play the roles that we've identified for UNDERSTANDING (again, with the caveat that some UNDERSTANDING-based successor would likely do even better). This leaves us without an easy verdict on Refocusing—it seems that we'll need *both* a KNOWLEDGE-like and an UNDERSTANDING-like concept to do all the epistemic work that needs doing in our conceptual practices. In the final section, then, we'll need to evaluate the relative importance of the roles themselves in order to draw even tentative conclusions about the best distribution of philosophical labour in epistemology.

For simplicity's sake, I'll call the roles we identified in Sect. 2.2 the 'KNOWLEDGE-roles' and the 'UNDERSTANDING-roles', respectively—though we ought to keep in mind that we're only claiming that KNOWLEDGE and UNDERSTANDING *prima facie* appear to serve these roles. Let's begin, then, by considering why UNDERSTANDING is poorly suited to the first KNOWLEDGE-role we identified, that of regulating permissible assertion. Here we immediately run into a complication—"understanding" is a highly polysemous word, and the different senses of this word seem differently suited to fill the KNOWLEDGE-roles. Particularly relevant for the regulation of assertion is the fact that certain senses of "understand" do not take propositional contents.

Following Kvanvig (2003), we can distinguish between 'objectual' and 'propositional' uses of "understanding". 'Objectual' uses involve an object or subject matter as content, as in "Sarah understands quantum mechanics". As noted earlier, understanding is typically supposed to involve something like a grasping of connections, structure, or something of the like within a body of information; as Grimm (2012) notes, it is "directed at a complex of some kind – in particular, at a complex with parts or elements that depend upon, and relate to, one another" (Grimm, 2012, p. 105). The objectual use of 'understand' fits naturally with this idea; to understand quantum mechanics is to understand a complex body of information and the interrelations among its parts. But, of course, objectual understanding is poorly suited to regulating assertion, for the contents of assertion are *propositions* rather than *objects*, and it is highly implausible that understanding an object is required before asserting propositions about that object. As an obvious example, surely it is possible to both permissibly and truly assert "I don't understand quantum mechanics".[19]

This suggests that we should look to propositional understanding; but here, we face the problem that the range of propositions that can be the contents of understanding seems much narrower than the range of propositions that can be asserted. Perhaps the most common sort of attribution of propositional understanding would be a claim like "Sarah understands that tides are caused by the gravitational pull of the moon". This sort of claim would typically express something more than mere knowledge of that proposition; it would imply that Sarah grasps or is able to explain the relevant causal connection between the tides and the moon. But this usage seems to only work for propositions that express causal connections. There are, admittedly, certain somewhat idiomatic propositional usages in English like "I understand that you're in the market for a new car" (in which case 'understand that' seems to mean 'have heard that'), or "I understand that you were hurt by my actions" (which seems to imply an emotional or empathetic state in addition to mere knowledge, but not necessarily any 'grasping of connections'). But these uses don't seem to get at the sort of state that fans of Refocusing are interested in. For most other propositions p, "S understands that p" is either unnatural (at least in English), or can at best be parsed as an alternative way of saying that S knows that P. Consider the sentence "Sarah understands that Paris is in France". The sentence itself is awkward, but moreover it's hard to make sense of what this could mean beyond mere knowledge attribution. What would the relevant 'complex' be? What sort of connections or relations could Sarah be grasping here? Propositions that involve causation or explanation, then, seem to be reasonable objects of propositional understanding—but such propositions form only a small proportion of assertable propositions, and much of the remainder doesn't appear to naturally admit of understanding. By contrast, just about every assertable proposition seems in principle knowable (outside of the occasional paradox-producing case, that is).

Even supposing that we could make sense of understanding that Paris is in France in some way that is not merely synonymous with knowing that Paris is in France, a

[19] A third type of understanding might be 'understanding-why', as in "I understand why the car won't start". It's an open question how this variety of understanding relates to the other two. However, understanding-why clearly is a non-starter for regulation of assertion, for the simple reason that one can clearly permissibly assert that p without knowing why p is the case.

more substantive obstacle remains. As we've seen, understanding is typically taken to be a fairly demanding cognitive achievement; it requires not only holding a belief on the basis of evidence, but also grasping certain relationships or connections within a body of information. Of course, some instances of understanding are easier to attain than some instances of knowing—for example, it is easier to come to understand why everyone is carrying umbrellas on a rainy day than it is to come to know whether there is sentient life beyond our solar system. But plausibly, for at least the vast majority of cases in which it is possible both to know and to understand some proposition p, understanding that p requires the greater cognitive effort. Indeed in many cases, understanding takes substantial time, effort, and ability. If a significant proportion of those 'effortful' cases involve propositions that we might plausibly wish to communicate, then this leaves open the possibility that understanding is simply too demanding a requirement to place on assertability.

In fact, I think there's a fairly general argument that understanding is an overly demanding requirement for permissible assertion. Consider the previously-noted fact that a large proportion of our beliefs come from testimony. Often, testimony occurs in chains—A tells something to B, and B passes on that information to C. But it's a commonly held view that understanding is very difficult, if not impossible, to pass along through mere testimony. Zagzebski, for instance, claims that "unlike beliefs, understanding is not passed along from a testifier to a recipient. The person's own mind has to do the "work" of understanding" (Zagzebski, 2009, pp. 145–146). Hills (2016), meanwhile, claims that 'understanding-why' requires a form of 'cognitive control' that normally cannot be passed on via testimony. If these authors are right, then limiting the assertable to the understood puts a very substantial barrier in the way of effective chains of testimony: an intermediate 'link' in the chain cannot simply acquire knowledge from A's assertion and then (permissibly) pass this along via assertion to C. She would first have to put in the additional work needed to come to understand the information she is passing along. It's not clear that this is always possible, or even desirable—even when the intermediary possesses the ability to understand the information she is passing along, it's not always worth the extra cognitive effort to do so.

The exact same considerations that make UNDERSTANDING a poor candidate for the norm of assertion suggest that UNDERSTANDING would serve poorly in the role of flagging appropriate sources of testimony. For most of our day-to-day testimonial needs, we don't need to find understanders. Knowers are plenty sufficient. This is partly linked to the fact that, in most cases, we *ourselves* don't need understanding from our testimonial sources. If I ask if anyone knows what time the meeting is, I'm not seeking someone who can enable me to grasp whatever structure or explanatory relationships might be needed in order to *understand* the meeting time (if indeed that even makes sense!) I just need to know when to show up, and thus I need to find a knower who can pass along that knowledge.

As the example above indicates, similar considerations can be brought to bear on the idea that UNDERSTANDING might serve in the role of regulating permissible action or practical reasoning. Again, even leaving aside the plausibly narrow range of propositions that are potential targets for understanding, it simply seems absurdly overdemanding to require understanding as a prerequisite for action. Let's use as an example a case that actually does plausibly admit propositional understanding: suppose I go to the doctor's office complaining of headaches and shortness of breath, and suppose that my doctor examines me and then informs me that my symptoms are caused by high blood pressure. Must I *understand* this causal connection before using it as a premise in my practical reasoning? I should hope not; advice from my doctor, alone, should be enough in this case to license taking action to lower my blood pressure. Aiming to understand before one acts may be admirable in certain cases, but it is typically not required. Moreover, given the difficulty of understanding, refraining from action until one understands is potentially dangerous, especially in cases where time is limited. An ideal epistemic concept for regulating action should plausibly pick out a state that is much less difficult to acquire; plausibly, about as difficult as knowing.

Our final 'epistemic role' was providing a stopping point for inquiry. Pritchard (2008) in fact claims that understanding, rather than knowing, is the aim of inquiry; part of his reasoning here relies on the claim that the goal of inquiry is typically to attain an answer to a 'why'-question. He then suggests that a person who ends inquiry after *merely* coming to know why, without understanding why, ends inquiry too soon. I admit to being puzzled by both of the claims involved in Pritchard's reasoning here. Certainly *some* inquiries aim at the answer to 'why'-questions, but it's not remotely clear that most do. Where are we meeting? How do I get to the train station? When is the next flight? What is the best-selling novel of 2018? It seems to me that the *overwhelming* majority of everyday inquiries do not concern 'why'-questions at all. And as before, the answers to many of the example questions just given don't obviously admit of understanding—it's not clear what it would mean, for instance, to understand that the next flight leaves at 10:26 (again, assuming that understanding this proposition is to mean something more than merely knowing it). In cases where the answers to these questions do potentially admit of understanding, moreover, it's not at all obvious that the average inquirer should be expected to attain understanding before terminating inquiry. For instance, perhaps one could understand that three cups of flour are needed for a particular recipe, in virtue of having a grasp of (e.g.) how the proportion of flour to sugar will affect the texture of the resulting cake. But the average home chef doesn't need to seek out this understanding; an understanding of the chemistry of baking is supererogatory, surely.[20]

What about 'why'-questions? I'd argue that even here, in most ordinary circumstances understanding is not required before permissible closure of inquiry. Often, we inquire into the answer to a 'why'-question because the answer is needed in

[20] One of the editors notes that the fan of Refocusing might appeal to a more restricted notion of inquiry-stoppage—for instance, they might claim that understanding is the appropriate stopping-point for *scientific* inquiry. This strikes me as a plausible move, though I couldn't say with certainty whether this is what e.g. Pritchard has in mind. In any case, I would suggest that even if understanding is well-suited to serve as an inquiry-stopper in science, there is a further need for some epistemic concept to play the role of a more general (and likely less stringent) standard to which ordinary inquiry is held.

order to guide our actions. And as we've already discussed, in the typical case understanding is not required for action. Consider a scenario where my computer has been crashing whenever I try to load up the media player. I might take to Google to search for an answer to the relevant 'why'-question (that is, 'why does my computer crash when I load the media player?'); after a short search, I might discover that the media player is crashing because of outdated drivers. I might then proceed to update the drivers and get on with watching pirated episodes of the Sopranos. Now, I'll admit to my own technical naïveté here—I don't even know what a driver is, really. But I do know how to update drivers for various programs on my computer, and I've more than once engaged in roughly the sort of inquiry just described. I certainly don't understand why outdated drivers might cause a crash; nonetheless, it strikes me as perfectly permissible to end inquiry without attaining said understanding. I've got shows to watch, after all.

There's a clear moral suggested by the observations we've made thus far. UNDERSTANDING picks out a mental state that is, compared to knowledge, relatively uncommon, both because of its restricted range of objects and because of the fairly demanding degree of cognitive achievement that it frequently requires. An enormous range of everyday contexts, however, call for use of an epistemic concept which picks out an epistemic standing that is both applicable to (nearly) all propositions, and that is relatively easy to attain.[21] KNOWLEDGE possesses these features; and though it may not be the absolute optimal concept to employ in such contexts, it certainly seems to outperform UNDERSTANDING. As noted earlier, that plausibly gives KNOWLEDGE a good claim to our theoretical attentions—at the very least, to employ it as a guide or model for an improved, custom-made concept to play the roles we've been discussing.

Now, what of KNOWLEDGE's suitability to fill the UNDERSTANDING-roles? The first role we discussed was that of identifying problem-solvers, particularly for novel or unexpected situations. There seems to be a good case to be made for UNDERSTANDING's superiority over KNOWLEDGE for this purpose. Understanding's link with the ability to solve novel problems, or to apply information to novel cases, seems to be partly constitutive of it—not so for knowledge. Note

[21] Again, as a reviewer has pointed out to me, there are a wide variety of cases where attaining understanding is intuitively not particularly difficult—as with, e.g., coming to understand why one's car won't start after noticing that the fuel meter reads empty. There is a bit of a complication here, since understanding is commonly held to come in degrees, and many such cases might fall on the lower end of the understanding spectrum. For instance, to *fully* understand why one's car won't start in the absence of fuel, one would presumably need to know quite a lot about how e.g. fuel is combusted and the energy produced thereby transferred to mechanical motion in the engine. The gradability of understanding itself arguably presents an obstacle to its straightforward employment as a threshold for permissible action, assertion, etc. (how *much* understanding is required for assertion?). But leaving this aside, we've noted previously that it's plausible that for a typical proposition p, understanding that p requires at least as much effort as knowing that p, and usually substantially more when 'higher' grades of understanding are involved. This increased demandingness doesn't seem to bring with it any increase in UNDERSTANDING's ability to fill the regulatory roles we've been discussing; and in many cases the increase in required effort/ability will be detrimental, as in several of the examples already given.

for instance Alison Hills' claim that moral understanding involves an ability to "draw the right conclusion or give the right explanation in new cases" (Hills, 2009, p. 102). Imagine a child who could report that it was wrong to take his little sister's toy, but who would return only a puzzled look when asked whether it was similarly wrong to take his brother's candy. Such a child might know that taking the toy is wrong, but would clearly fail to understand why. This is not peculiar to ethical understanding, either. Consider how you might design a test to evaluate a students' understanding of a particular theory (either philosophical or otherwise). Likely, you would aim to produce a novel scenario and ask the student how the theory applies in the new case. A student who can report that logical positivists deemed 'God exists' to be meaningless but isn't so sure what they'd say about 'red things are red in virtue of participation in the Form of Redness' likely doesn't fully understand logical positivism (or perhaps Platonic realism). Nonetheless, that student might accurately be said to know. This suggests, I'd argue, that knowledge is too weak to serve well in the role of identifying problem-solvers.

What about the role of characterizing scientific inquiry? Alexander Bird (2007) argues that knowledge is in fact the aim of scientific inquiry—though he contrasts this view with one according to which science merely aims at the accumulation of true belief. On the other hand, the sorts of features that distinguish understanding from mere knowing—grasping of connections/pattern, etc.—seem like just the sort of features science aims to uncover. Understanding's close link with explanation, moreover, plausibly suggests that UNDERSTANDING might be more suited to this role than KNOWLEDGE. Finally, the aforementioned relatively high cognitive demandingness of understanding is no real barrier in a scientific context—indeed, it seems appropriate to demand high-quality epistemic states in a context where participants expend abnormally large quantities of time, cognitive energy, and government funding on inquiry. The case here seems to me to be rather less conclusive than the case against using UNDERSTANDING for the KNOWLEDGE-roles; however, I'm moderately inclined to give the palm to UNDERSTANDING. More generally, I suspect that whatever concept best fills the UNDERSTANDING-roles will be that of a somewhat 'higher' epistemic state than knowledge, thus positioning UNDERSTANDING to serve as a model for future conceptual tinkering.

2.4 Drawing Conclusions

Let's sum up. There seem to be a variety of cognitive roles for which our current concept KNOWLEDGE is plausibly used. KNOWLEDGE may not be the ideal concept to fill these roles, but UNDERSTANDING is distinctly unsuited to the task. On the other side of the coin, there are a few cognitive roles that do seem to be better suited to the deeper, more demanding sort of epistemological status expressed by UNDERSTANDING—though here too a custom-made successor might serve even better. What ought we to conclude with regard to the Refocusing thesis?

To really draw a careful conclusion here, we would need to supplement our conceptual engineering investigations with a set of normative claims about the most desirable distribution of cognitive labour within epistemology. I've been able to provide only the most basic sketch of such claims in this chapter—as noted earlier, I find it plausible that the distribution of cognitive labour should be in large part guided by the relative values of the *concepts* under consideration (rather than their objects). In turn, I suggested that these values would be in some way or another a function of the roles these concepts play in our cognitive lives. So given that both KNOWLEDGE-like and UNDERSTANDING-like concepts seem to be needed to fully fill the various roles we've been discussing, what can be said about the relative importance of these roles?

I'll offer up just a few brief arguments to the effect that a concept that effectively fills the KNOWLEDGE-roles is to some degree more theoretically valuable than a concept that effectively fills the UNDERSTANDING-roles. First, we might find some weak support from speculating on the quantity of said roles. As noted earlier, there are several roles played by KNOWLEDGE that we did not even have space to carefully examine, such as its role in prediction and explanation of behaviour, and in the assignment of moral blame. I suspect there might be more still. A second, more substantive argument might appeal to the ubiquity of the contexts in which the KNOWLEDGE-roles are active. Assessments of assertion, action, testimony and inquiry are nigh-constant features of our day-to-day interactions with the world. Scientific activity and the recruitment of experts for problem-solving purposes are clearly useful as well, but not nearly so omnipresent.

On the other hand, one might argue that the contexts in which the UNDERSTANDING-roles are active are in some sense more 'important', despite their relative rarity. This might be cashed out in terms of the payoff that would result from improving our theories of understanding, or indeed of improving the concept UNDERSTANDING itself—improvements to scientific methodology, perhaps. Nonetheless, the payoff for devoting resources towards the study of KNOWLEDGE also promises to be significant; consider the practical benefits of an ability to precisely identify the appropriate epistemic cutoff for acting, asserting, and closing inquiry!

A final consideration might be found in the idea that knowledge or knowledge* is plausibly a more fundamental or basic epistemic state than understanding/understanding*, in the sense that it might not be possible to understand(*) anything at all without first possessing quite a bit of knowledge(*).[22,23] If so, a solid theory of

[22] This is not to claim that understanding(*) is reducible to knowledge(*) or even that one cannot understand(*) p without knowing(*) that p. It is instead to claim that one cannot understand(*) something—let's say, the proposition that the tides are caused by the moon—without being in possession of a variety of pieces of knowledge (knowledge that the moon exerts gravitational pull, e.g.) that at least partially ground one's 'grasping' of the interconnections/relations/etc. that enable understanding.

[23] I'm indebted to Weng Hong Tang, Davide Fassio, and Jie Gao for comments on a presented version of this chapter which inspired this last point.

knowledge(*) might be a prerequisite to a complete theory of understanding(*). That seems to me to weigh at least a little in favor of the importance of KNOWLEDGE(*).

Though there's much more that could be said, space considerations prevent me from going further without risking the wrath of the editors. I'll thus have to rest content with *very* tentatively endorsing the following weak version of Refocusing: UNDERSTANDING and its potential successor-concepts currently receive too little theoretical attention, but a KNOWLEDGE-like concept nonetheless plausibly deserves the lion's share of our theoretical attention in epistemology. Ultimately, however, I'm less concerned with convincing the reader of any particular version of Refocusing, and more concerned with motivating the overall approach. When deciding where to focus our epistemological attentions, conceptual engineering provides a novel perspective by emphasizing the purposes to which our epistemic concepts are put. What's more, conceptual engineering reminds us that existing concepts shouldn't exhaust our options—we've thought too little about UNDERSTANDING, but there are yet other epistemic concepts that we haven't thought about at all.

References

Bird, A. (2007). What is scientific progress? *Noûs, 41*(1), 64–89.

Bourget, D. (2017). The role of Consciousness in grasping and understanding. *Philosophy and Phenomenological Research, 95*(2), 285–318.

Brown, J. (2010). Knowledge and assertion. *Philosophy and Phenomenological Research, 81*(3), 549–566.

Cappelen, H. (2018). Fixing Language: *An Essay on Conceptual Engineering*. Oxford University Press.

Craig, E. (1990). *Knowledge and the state of nature: An essay in conceptual synthesis*. Clarendon Press.

de Regt, H. W. (2009). The epistemic value of understanding. *Philosophy of Science, 76*(5), 585–597.

DeRose, K. (2002). Assertion, knowledge, and context. *The Philosophical Review, 111*(2), 167–203.

Elgin, C. (2006). From knowledge to understanding. In S. Hetherington (Ed.), *Epistemology futures* (pp. 199–215). Oxford University Press.

Fantl, J., & McGrath, M. (2002). Evidence, pragmatics, and justification. *The Philosophical Review, 111*(1), 67–94.

Fassio, D., & McKenna, R. (2015). Revisionary epistemology. *Inquiry, 58*(7–8), 755–779.

Gettier, E. (1963). Is Justified True Belief Knowledge? *Analysis, 23*(6), 121–123.

Grimm, S. (2012). The value of understanding. *Philosophy Compass, 7*(2), 103–117.

Hannon, M. (2019). *What's the point of knowledge? A function-first epistemology*. Oxford University Press.

Hannon, M. (2021). Recent work in the epistemology of understanding. *American Philosophical Quarterly, 58*(3), 269–290.

Hansen, N., Porter, J. D., & Francis, K. (2021). A corpus study of "know": On the verification of philosophers' frequency claims about language. *Episteme, 18*(2), 242–268.

Haslanger, S. (2020). How not to change the subject. In T. Marques & A. Wikforss (Eds.), *Shifting concepts: The philosophy and psychology of conceptual variation* (pp. 235–259). Oxford University Press.

Hawthorne, J. (2004). *Knowledge and lotteries*. Oxford University Press.

Hawthorne, J., & Stanley, J. (2008). Knowledge and action. *The Journal of Philosophy, 105*(10), 571–590.

Hempel, C. (1965). *Aspects of scientific explanation*. Free Press.

Hills, A. (2009). Moral testimony and moral epistemology. *Ethics, 120*(1), 94–127.

Hills, A. (2016). Understanding why. *Nous, 50*(4), 661–688.

Kappel, K. (2010). On saying that someone knows: Themes from Craig. In A. Haddock, A. Millar, & D. Pritchard (Eds.), *Social epistemology* (pp. 69–88). Oxford University Press.

Kelp, C. (2011). What's the point of "knowledge" anyway? *Episteme, 8*(1), 53–66.

Khalifa, K. (2012). Inaugurating understanding or repackaging explanation? *Philosophy of Science, 79*(1), 15–37.

Kvanvig, J. L. (2003). *The value of knowledge and the pursuit of understanding*. Cambridge University Press.

Kvanvig, J. (2009). The value of understanding. In A. Haddock, A. Millar, & D. Pritchard (Eds.), *Epistemic value* (pp. 95–111). Oxford University Press.

Lackey, J. (2007). Why we don't deserve credit for everything we know. *Synthese, 158*(3), 345–361.

Machery, E. (2020). What is the function of the concept of knowledge? [Video]. *YouTube*. https://www.youtube.com/watch?v=HMPlnb-WWGM

Prinzing, M. (2018). The revisionist's rubric: Conceptual engineering and the discontinuity objection. *Inquiry, 61*(8), 854–880.

Pritchard, D. (2007). Recent work on epistemic value. *American Philosophical Quarterly, 44*(2), 85–110.

Pritchard, D. (2008). Knowing the answer, understanding and epistemic value. *Grazer Philosophische Studien, 77*(1), 325–339.

Pritchard, D. (2009). Knowledge, understanding and epistemic value. *Royal Institute of Philosophy Supplement, 64*, 19–43.

Pritchard, D. (2014). Knowledge and understanding. In A. Fairweather (Ed.), *Virtue epistemology naturalized: Bridges between virtue epistemology and philosophy of science* (pp. 315–329). Springer International Publishing.

Riggs, W. (2003). Understanding 'virtue' and the virtue of understanding. In M. De Paul & L. Zagzebski (Eds.), *Intellectual virtue: Perspectives from ethics and epistemology* (pp. 203–226). Oxford University Press.

Rysiew, P. (2012). Epistemic scorekeeping. In J. Brown & M. Gerken (Eds.), *Knowledge ascriptions* (pp. 270–293). Oxford University Press.

Simion, M., & Kelp, C. (2020). Conceptual innovation, function first. *Noûs, 54*(4), 985–1002.

Stanley, J. (2005). *Knowledge and practical interests*. Oxford University Press.

Thomasson, A. L. (2020). A pragmatic method for conceptual ethics. In H. Cappelen, D. Plunkett, & A. Burgess (Eds.), *Conceptual ethics and conceptual engineering* (pp. 435–458). Oxford University Press.

Trout, J. D. (2002). Scientific explanation and the sense of understanding. *Philosophy of Science, 69*(2), 212–233.

Turri, J. (2015). Understanding and the norm of explanation. *Philosophia, 43*(4), 1171–1175.

Wilkenfeld, D. A., Plunkett, D., & Lombrozo, T. (2016). Depth and deference: When and why we attribute understanding. *Philosophical Studies, 173*(2), 373–393.

Williamson, T. (1996). Knowing and asserting. *The Philosophical Review, 105*(4), 489–523.

Williamson, T. (2000). *Knowledge and its limits*. Oxford University Press.

Zagzebski, L. (1994). The inescapability of Gettier problems. *The Philosophical Quarterly, 44*(174), 65–73.

Zagzebski, L. (2001). Recovering understanding. In M. Steup (Ed.), *Knowledge, truth, and duty: Essays on epistemic justification, responsibility, and virtue* (pp. 235–252). Oxford University Press.

Zagzebski, L. (2009). *On epistemology*. Wadsworth.

Chapter 3
Engineering Evidence

Mona Simion

3.1 Introduction

Evidence matters: the concept of evidence is central to epistemology, the philosophy of science, philosophy of law, the ethics of responsibility. Outside philosophy, the concept of evidence is highly employed as well: lawyers, judges, historians and scientists, economists, investigative journalists, and reporters, as well as ordinary folk in the course of everyday life talk and think about evidence a lot.

Both within and outwith philosophy, what we care most about is not just the nature of evidence alone, but rather about what it is for a subject to *have* evidence. We care, as it were, about evidence had. That makes sense in philosophy, because we are interested in the quality of our beliefs and our actions, and the latter will mostly be affected by the evidence we have. Outwith philosophy, evidence one has bears relevance to one's legal status, professional performance, decisions, policies, voting, plans.... Evidence one has, the thought goes, but less so evidence that one does not have, will influence all of one's walks of life.

It is interesting to note that the ways in which we think of evidence one has outwith and within philosophy are strongly incompatible with each other (Kelly, 2016): in philosophy we disagree a lot about the nature of evidence, but one thing the vast majority of theorists have always assumed is that the having relation is somehow related to the limits of one's skull: one has evidence, on this received view, when one uptakes it 'in one's head'—be it via seemings, beliefs, knowings etc. (Kelly, 2016). In contrast, outwith philosophy, the having relation has never been about the skull: just try to tell a judge that you had no evidence that the butler did it, even though he did it right in front of you, because you couldn't believe your eyes; see how that goes down.

M. Simion (✉)
Cogito Epistemology Research Centre, University of Glasgow, Glasgow, Scotland, UK
e-mail: mona.simion@glasgow.ac.uk

M. G. Isaac et al. (eds.), *New Perspectives on Conceptual Engineering - Volume 2*, Synthese Library 514, https://doi.org/10.1007/978-3-031-98524-9_3

Of course, one might think, what is the surprise there? Experts know best in all domains—that is what semantic externalism teaches—and philosophy is not an exception. The way in which us laymen conceptualise 'depression' is likely different from the way in which psychiatrists do; that's fine, we're wrong, and the experts are right. The same goes for evidence one has.

In what follows, I will argue that, in the case at hand, experts are wrong: the philosophical conception of the having relation—as having to do with the limits of one's skull—fails on extensional grounds, and, as a result, it also fails to fulfil its central function in predicting accountability, and thereby loses its legal significance. Furthermore, I will argue we should adopt my preferred conception of the having relation: On this account, having evidence has to do with being in a position to know.

The paper proceeds as follows: first, I show that the vast majority of extant accounts of the having relation suffer from an in-principle theoretical problem: they cannot accommodate cases of evidence and defeat resistance—such as climate change denialism or vaccine hesitancy in the face of overwhelming expert testimony (Sect. 3.2 and 3.3). This theoretical problem, in turn, renders these views very problematic for use outside of philosophy, where the interest in the having relation mainly has to do with holding people accountable for acting in disregard of available evidence, and for the consequences of the relevant actions. In Sect. 3.4, I propose that we should engineer a conception of having evidence that corrects for both these problematic features, and thus is both philosophically neat and useful outside of philosophy.

This is, in a first instance, a project in conception engineering: it aims to bring the expert conception of the having relation in line with the straightforward, laymen conception thereof as having to do with easy availability rather than the limits of one's skull.

Conceptions are not concepts: many of us have the concept of justice, but most of the theories of justice on the market will put forth the wrong conception thereof. Similarly, it is perfectly possible (and, I submit, plausible) that we (philosophers and laymen alike) host the right concept of evidence had (intuitions that something goes wrong in cases of evidence resistance do support this hypothesis), but philosophers traffic in the wrong conception thereof. In this, this is a project in conception engineering.

In a second instance, however, the project may also be one in conceptual engineering (depending on one's theoretical commitments): if we do turn out to host the wrong concept of evidence had, and if experts have causal power over meanings, expert conception change will help with meaning change.

3.2 Evidence Had

In previous work (Simion, 2023, 2024) I have discussed at length cases of resistance to evidence, and I have argued that extant accounts of the nature of evidence fail to do them justice. In what follows, I will argue that the culprit lies with the way in which extant accounts unpack the having relation. Consider the following cases of evidence resistance:

Case §1. Testimonial Injustice: Anna is an extremely reliable testifier and an expert in the geography of Glasgow. She tells George that Glasgow Central is to the right. George believes women are not to be trusted, therefore fails to form the corresponding belief.

Case §2. Unwarranted Optimism: Mary is an optimist. When her partner Dan spends more and more evening hours at the office, she's happy that his career is going so well. When he comes home smelling like floral perfume, she thinks to herself: 'wow, excellent taste in fragrance!' Finally, when she repeatedly sees him having coffee in town with his colleague Alice, she is glad he's making new friends.

Case §3. Misdirected Attention: Professor Racist is teaching College maths. He believes people of colour are less intelligent than white people. As a result, whenever he asks a question, his attention automatically goes to the white students, such that he doesn't even notice the black students who raise their hands. As a result, he believes black students are not very active in class.

What is epistemically wrong in these cases? Here is what my non-philosopher grandmother would say, off the top of her head: these people have overwhelming evidence available to them that p is the case (my grandmother uses p-s and q-s a lot), but they fail to believe p nevertheless. That is what is (epistemically) wrong with these cases: it's an update failure. Let's call this 'the easy answer': it's the answer that explains the epistemic failure at stake in the resistance cases above as involving the having relation.

Unfortunately, epistemologists don't have the luxury of the availability of the easy answer that my grandmother enjoys. That is because, according to both internalist and externalist theories of the having relation, the latter has to do with evidence being, in one way or another, within one's skull (Kelly, 2016).

3.2.1 Internalism About Having Evidence

Epistemic internalisms[1] (of all sorts) notably struggle with cases like the ones under discussion: according to these theories, having evidence has to do with the presence of a particular mental state—usually a seeming, or an experience of sorts. It is easy to see that this way to think of the having relation cannot explain what is going on, e.g., in the case of Professor Racist: after all, this persons's racist belief that people of colour are less intelligent than white people make it such that his attention automatically goes to the white students, and he doesn't even notice the black students

[1] See Pappas (2023) for an excellent overview of the debate between epistemic internalism and epistemic externalism, and (Kelly, 2016) for internalist and externalist views of evidence.

who raise their hands. As a result, he believes black students are not very active in class. Since he doesn't even notice the black students, Professor Racist is not undergoing the relevant seemings—indeed, he fails to instantiate any mental states whatsoever with the relevant evidential content (i.e. 'Black students raised their hands'). On seemings internalism, then, he has no evidence that black students raised their hands, so the explanation of the datum that Professor Racist's belief that black students are not active in class is intuitively unjustified remains mysterious.

Similarly, internalism will predict, against intuition, that Mary, the wishful thinker, has evidence to believe that her husband is faithful—after all, it seems to her that her husband is faithful.

This latter kind of problem is a classic one for internalism—it is a problem having to do with seemings with bad etiologies. It seems that what is needed in these cases is an external difference maker between good and bad etiologies, which internalism just doesn't have access to. Given this, in what follows, I will focus on what I take to be the most prominent externalisms on the market—i.e. reliabilisms and knowledge-first accounts—and argue that their rendering of the having relation fails to explain the impermissibility datum in the resistance cases.

3.2.2 *Evidence Had as Evidence Known*

According to the prominent, knowledge-first view of having evidence (Williamson, 2000), for any subject S, S's evidence is S's knowledge. Since knowledge implies belief, and since subjects like e.g. Professor Racist lack the relevant beliefs, E = K will predict that the subjects in question lack evidence. In this, the E = K champion does not have the easy answer available to them.[2]

Recent work by Maria Lasonen-Aarnio and Tim Williamson suggests the E = K theorist can make good on the resistance datum—i.e., the intuitive epistemic impermissibility present in resistance cases—via employing epistemic blameworthiness. Here is the view, in rough outline: an agent is blamelessly in breach of epistemic norms only if the breach in question occurs in spite of the fact that the subject in question does what a believer with good epistemic dispositions would do. Conversely, breaches of epistemic norms that are such that a believer with good epistemic dispositions would do otherwise render the subject in question epistemically blameworthy. George, for instance, in not believing what the woman passer-by tells him, fails to do what a good believer—i.e. a believer with good epistemic dispositions—would do, and, as such, is epistemically blameworthy.

Otherwise put, the thought is that what is present in Cases 1–3 and explains the impermissibility datum is not evidence the subjects in question have, but evidence their counterparts manifesting good epistemic dispositions *would have had.*

[2] Note that Cases 1 and 2 can also be redescribed as involving lack of any beliefs—including the lack of the belief 'the woman told me that p', or 'my husband is late once again' etc.

The main problem with this move, however, concerns extensional adequacy: one need not have bad epistemic dispositions in order to fail, epistemically, in the way in which resistant cognisers do: it can be a one-off affair. Maybe e.g. Mary the wishful thinker is an excellent believer otherwise, and in all other walks of life. Since being a good believer—i.e. having good epistemic dispositions—plausibly has to do with reliability rather than infallibility, Mary will count as having good epistemic dispositions. If this is the case, Mary does do what a believer with good epistemic dispositions—namely herself—would do in this situation. A believer with good epistemic dispositions, in virtue of the fallibility thereof, sometimes gets it wrong.

Furthermore, zooming in on the particular disposition involved in the formation exercise at stake will not do much to help either. Reliable dispositions, even if they are reliable for cases like the case under consideration, they need not be infallible in cases like the case under consideration. I submit that, no matter how we choose to individuate the disposition at stake, insofar as the theory demands mere reliability rather than infallibility, one-off failures are to be expected. Our theory, then, should be able to explain the intuition that something is wrong in these cases, in spite of the reliable disposition being present (and manifest). The question that remains unanswered by the blameworthiness response, then, is what normative category the impermissibility intuition in the resistance cases tracks to begin with.

3.2.3 Reliable Indicators

We have seen that knowledge-based externalisms about the having relation struggle to accommodate the resistance data. In what follows, I will look at reliabilist externalisms about evidence and reasons to believe, in search for the normative resources we need to this effect. In this section, I take on Juan Comesaña's (2010) indicator reliabilism. In the next section, I will look at virtue reliabilist (e.g. Sylvan & Sosa, 2018; Burge, 2013) accounts of reasons to believe.

According to Comesaña's account of evidence, an experience provides its content as a reason when the subject is justified in believing its content. The justification in question in the account, importantly, (1) is non-factive, and (2) must be *ultima facie*: if an experience of the subject S provides them with *prima facie* justification for believing its content but this justification is defeated by something else S is justified in believing, then S does not have the content of the experience as evidence (2010, p. 119).

This raises the question: when are we *ultima facie* justified in believing the contents of our experience? The answer to this question will tell us how Comesaña's account fares on extensional adequacy.

Comesaña favours a reliabilist account of justification. When coupled with his view of evidence, this will render the latter better positioned to distinguish good from bad experience etiologies than internalism about evidence—in that only those contents of experience that are believed based on a reliable process will qualify as evidence.

Indeed, at first glance, Comesaña's account promises to deliver the result we want in several of the resistance cases. Take, for one, Testimonial Injustice: Since on Comesaña's view an experience provides its content as a reason just in case the subject is *ultima facie* justified in believing its content, and since, by stipulation, Anna is an extremely reliable testifier, the content of George's experience as of her telling him that Glasgow Central is to the right constitutes evidence. Anna's testimony provides George with a reason to believe Glasgow Central is to the right, which he fails to take up. Similarly, Mary has evidence that her husband is cheating, that Dave, the detective has evidence that the butler did it, etc. All of these people have experiences with the relevant contents that are reliably generated, the contents of which thus count as evidence on Comesaña's view.

Unfortunately, on closer look, it turns out that granting the indicator reliabilist success on the resistance front is a bit fast: just like seemings internalism, indicator reliabilism gets cases that do not feature experiences wrong. The view will predict that Racist has no evidence that the black students are active in class. After all, he has no experience with this particular content, therefore no reliably generated experience with this particular content either. Furthermore, note that the case of Professor Racist is not unique in this respect: we can modify all the other cases along the exact same lines—i.e. ramping up the epistemically bad features—to get the same wrong predictions. Here is how: First, we can make it such that our characters not only don't form the relevant beliefs because of sexism, politically motivated reasoning etc., but they don't even host the corresponding experiences. Say that George, for instance, in Testimonial Injustice, not only doesn't believe what Ann says, but he doesn't even register that she said anything at all, due to his sexist bias. In all cases like these, indicator reliabilism will predict absence of had evidence.

3.2.4 Virtue Reliabilist Reasons to Believe

Virtue reliabilist accounts of evidence are, to my knowledge, missing in the literature. However, the market features well-developed virtue-theoretic views of had reasons to believe (Sylvan & Sosa, 2018; Burge, 2013). On a view e.g. equating evidence with epistemic reasons to believe, the epistemic reasons having relation will coincide with the evidence having relation.

According to these authors, broadly speaking, competences come first in epistemic normativity: a fact is S's epistemic reason to believe for S just in case it is competently taken up and processed by S. At root, then, reliable epistemic competence is doing the work in accounting for the having relation: "[We] think […] claims [about reasons supporting a species of justified belief] could only be true if possession and proper basing are themselves grounded in a deeper normative property of competence" (Sylvan & Sosa, 2018).

In turn, epistemic competences are traditionally unpacked as dispositions to believe truly (Sosa, 2016), or know (Kelp, 2018; Miracchi, 2015; Schellenberg, 2018).

The view, whether construed along truth-first or knowledge-first lines, is too weak to account for what is going wrong in cases of resistance to evidence: Think back to the case of George, the sexist; on this view, we get the result that George has no reasons to believe what the woman passer-by tells him, since he is not uptaking the relevant facts via his cognitive competences. The same will hold for all Cases 1–3: All these facts fail to constitute had epistemic reasons on this view, since they are not competently processed by the subjects.

But can't the virtue theorist appeal to these epistemic agents' lack of competence to explain the poor epistemic status of beliefs that they do hold, and account thereby for the impermissibility of resistance intuition?[3] For instance, can't the virtue theorist argue that what is going on in cases like e.g. Testimonial Injustice is that George is an epistemically incompetent believer in virtue of his sexist bias? This, the virtue theorist may argue, is enough to explain the intuition of epistemic permissibility.

Two things about this: first, just like in the case of the blameworthiness move, note that it need not be that the subject at stake is a rubbish epistemic agent overall. Indeed, again, maybe Mary, the wishful thinker, is actually an extremely reliable believer. It's only on this particular instance that she gets it wrong (after all, competences need not be infallible, but merely reliable; thereby, their presence is compatible with occasional failures). If so, the virtue theorist cannot appeal to lack of competence to explain this datum.

One way to go for the virtue theorist here would be to blame the impermissibility on impermissible suspension. In recent work, Sosa proposes to account for what is going wrong in cases such as these by making use of a notion of epistemic negligence: he suggests that negligent failure to inquire may preclude competent judgement. If so, we could maybe avail ourselves of this normative resource to explain how negligent failure to inquire may preclude competent suspension as well. Sosa's key idea is that if you can assess your first order competence by more reliable means but fail to do so, then you are negligent. In particular, you fall foul of a kind of negligence that precludes what he calls the competent attainment of aptness.

Most importantly for present purposes, given that competent suspension requires that one suspends based sufficiently on one's lack of the competence required in order to answer the question aptly, negligence precludes competent suspension: sexist George does have sufficiently available means that would enable a more reliable assessment of the aptness of his suspension—Anna's testimony. Since he ignores it, George will count as a negligent suspender. (By now, the details of this account of negligence should sound very familiar to the reader from back when we discussed the Alternative Reliable Process account of defeat).

The problem with this account of negligence is that it makes negligence too easy to come by. To see this, consider a case in which I ask my flatmate who is currently in the kitchen whether we have any milk left. He tells me that we do. Now, I do have several more reliable means of assessing my first order competence available to me. For instance, I could go to the kitchen and have a look myself. Crucially, however,

[3] Many thanks to Ernie Sosa for suggesting this route to me.

failure to avail myself of these means doesn't make me negligent. And, most importantly for present purposes, it doesn't preclude my judgement that there is milk in the fridge from being competent.

Sosa's account of negligence is insufficiently normative. What matters, according to Sosa, is the *availability of alternative means* that would lead to a more reliable assessment of first order aptness and competence. However, the difference maker is normative, not descriptive: what matters is not (only) whether one has alternative means available that would have led one to a more reliable assessment of first order aptness and competence, but (also) whether one *should* have availed oneself of these means. In the case of sexist George, he should have taken the woman's testimony into account in making up his mind on the location of Glasgow Central. Similarly, e.g. Mary should not have ignored all the evidence suggesting that her husband is having an affair. In contrast, in the milk case, it is not the case that I should have had a look myself.

3.3 Evidence One Should Have Had

The above results suggest that one way to attempt to explain the datum in resistance cases is by arguing that what is at stake in these cases is not evidence had, but rather evidence one *should* have had.

One such view in the literature is championed by philosophers who think that resistance cases are not cases of genuinely epistemic failure, or rather cases of moral failure. On this account of the data, the intuition of impropriety in the resistance cases has a non-epistemic normative source: we think, for instance, that George is doing something wrong in the Testimonial Injustice case because he's doing something morally wrong in not listening to the female passer-by: the testimony form the female passer-by is not evidence he has, but is evidence he (morally) should have had. Epistemic injustice, the thought would go, is the stuff of intellectual ethics, not of theory of knowledge proper. However, our intuitions are not fine grained enough to see the difference: theory is needed. Indeed, here is Richard Feldman:

> It's surely true that there are times when one would be best off finding new evidence. But this always turns on what options one has, what one cares about, and other non-epistemic factors. As I see it, these are prudential or moral matters, not strictly epistemic matters. (2008, p. 190)

I don't find this move particularly plausible: the failure in question in Cases 1–3 is a genuinely epistemic failure (compatibly, of course—and plausibly—it is also a moral failure). To see this, note that, while some of these cases exhibit ethically problematic features (e.g. Case #1, Case #3), Case #2 does not. To the contrary, Case #2, can even be construed as a case of moral success: placing more trust than the evidence supports in a member of one's family has been long discussed in the literature on partiality in friendship as morally permissible, if not even mandatory. It's hard to find moral flaws with Mary's epistemic ways. This suggest that the source of the intuition is, indeed, epistemic failure (absent other normative constraints at the context).

In recent work, Sandy Goldberg has taken up the task of developing an account of the normativity of evidence one *should have had* that vindicates the thought that what is at stake in these cases is epistemic failure—i.e., whereby the should in question is an epistemic rather than a moral should—, but that grounds the epistemic failure at stake in social normativity. One key thought that motivates Goldberg's project is that social roles—for instance, being a medical doctor—come with normative expectations. These normative expectations may be, and often enough are, epistemic. For instance, there is a social epistemic expectation that medical doctors are up speed on relevant literature in their field. Another key thought is that to believe that *p* justifiably one must live up to all of these legitimate expectations. Doctors who fail to be up to speed with the most recent research in their field are not justified in their corresponding beliefs, in virtue of being in breach of the social expectation associated with their role. For instance, a doctor who believes that ulcer is caused by stress, in ignorance of the widely available evidence that it is caused by bacteria, is not justified to believe that ulcer is caused by stress. As such, Goldberg grounds the normativity of evidence one should have had in the social expectations associated with the believer in question's social role.

It is easy to see that Goldberg's key thoughts promise to give us the ideal resources to handle cases of resistance to evidence. Take, for instance, the case of Professor Racist. Recall that this fellow is biased against people of colour, and as a result, whenever he asks a question, his attention automatically goes to the white students, such that he doesn't even notice the black students who raise their hands. In virtue of occupying his social role, by the first key thought, Professor Racist is subject to normative expectations that are associated with this social role. In particular, he is subject to the expectation to fairly distribute his care and attention in the student population. Since Professor Racist doesn't live up to this expectation, by the second key thought, he does not believe justifiably that the black students are not active in his class.

This is a very rough description of how Goldberg aims to deal with the kind of case of resistance to evidence that we are concerned with here. Even so, here is a worry that arises immediately: social expectations can be legitimate social expectations, but also, illegitimate social expectations. Women, for instance, are often illegitimately expected to carry most of the household burden, and to underperform in leadership roles. If so, it would seem as though social expectations cannot play the normative grounding role that Goldberg wants them to play, since they seem to require further normative unpacking themselves: we seem to need further normative notions to help distinguish between legitimate and illegitimate social expectations. When it comes to epistemic normativity in particular, things aren't looking much better: consider societies in which women are considered to be unreliable testifiers because intellectually inferior: in these societies, hearers will be expected to disbelieve women's testimony. Clearly, this social expectation is epistemically illegitimate. If so, the question that arises is: Isn't the good variety of social epistemic expectations that Goldberg appeals to in order to explain intuitive epistemic failure grounded in epistemic norms? And if so, won't we have to invoke the relevant epistemic norms in the final analysis of what goes wrong in cases of resistance to evidence? As a result, Goldberg's story will struggle to offer a satisfactory account of resistance cases.

3.4 Having Evidence as Being in a Position to Know It

This section argues we—philosophers—should conceive of the having relation in the same way in which my grandmother does: as having to do with easy availability rather than the limits of one's skull.

Here are a few reasons to do so: first, such a conception should be preferred on counts of simplicity and extensional adequacy. Recall: being able to say that the subjects in the resistance cases have the relevant evidence has the major advantage of opening the way towards what I called 'the easy answer' concerning what goes epistemically wrong in the resistance cases: it's an update failure. These subjects have evidence that supports that p and fail to correspondingly update their beliefs.

Second, we have already seen that there are major difficulties involved in trying to account for the impermissibility datum in terms of evidence one should (morally, socially, epistemically) have had. If so, we have further reason to suspect we should get back to the drawing board, and start with the straightforward, folk intuition of what goes wrong in these cases (as per, e.g., Derose's (2006)) preferred 'methodology of the straightforward').

Third, note that a conception of the having relation that fails to account for resistance cases is fantastically mal-equipped for moral and practical societal functions, in virtue of its incapability to predict accountability. What evidence we have matters for the permissibility of our actions in all walks of life, and crucially in accountability-centric fields such as law, public communication, politics, healthcare, and so on.

The first and second reason for preferring the layman conception of the having relation suggest conception engineering is warranted on strict theoretical grounds: our current conception fails to reliably track the entity in the world. The last reason has to do with conceptual ethics: we should engineer a conception of the having relation that does the important moral, social, practical work that we need it to do.

In previous work (REDACTED 2018, 2020), I have defended a view according to which conceptual engineering (broadly construed—be it about concepts, concepts, intensions, words, etc.) is obligatory in cases in which there is all-things-considered reason to do so, and no epistemic loss is involved. In what follows, I will follow suit. I propose that, in line with folk conception of the having relation, we should think of evidence one has as evidence that one is in a position to know.

This account of the having relation is, in principle, compatible with a variety of views of the nature of evidence itself (facts, propositions), and the nature of evidential support (probabilistic, modal etc.). My preferred view (Simion, 2023, 2024) takes evidence for p to be facts that enhance one's evidential probability that p is the case. In turn, evidence one has, on this account, will consist in probability enhancers that one is in a position to know. Here is the account in full:

Evidence as Knowledge Indicators: a fact e is S's evidence for a proposition p just in case S is in a position to know e, and S's evidential probability that p is the case conditional on e is higher than S's unconditional evidential probability that p is the case.

The having relation has to do with being in a position to know. What is it for me to be in a position to know e? A certain availability relation needs to be instantiated. On my view (Simion, 2023, 2024), availability has little to do with the limits of my skull. Evidence may consist of facts 'in the head' or facts in the world. Some facts—whether they are in the head or in the world, it does not matter—are available to me, they are, as it were, 'at hand' in my (internal or external) epistemic environment. Some—whether in the head (think of justified implicit beliefs, for instance) or in the world, it does not matter—are not thus available to me.

My notion of availability will track a psychological 'can' for an average cognizer of the type at stake; generality problems aside, for the vast majority of us, the relevant type will be 'human adult cognizer': the 'can; at issue will track what someone with an average adult human cognitive architecture can do.

There are *qualitative* limitations on availability in this sense: we are cognitively limited creatures. There are of *types* information that we just cannot access, or process: the fact that there is a table in front of me is something that I can easily enough access. The fact that you're dancing in the flat next door is not something I can easily access: I can't see through walls. I am in a position to *come to* know it—should I walk down the hallway and ring your bell—but I am not in a position to know it, as things stand (henceforth also *Qualitative Availability*).

There are also *quantitative* limitations on my information accessing and processing: the fact that there's a table somewhere towards the periphery of my visual field—in contrast of it being right in front of me, in plain view—is not something I can easily process: I lack the power to process everything in my visual field, it's just too much information: easier availability gets epistemic priority. Similarly, while I might easily access facts f, f1, f2, f3… fn independently, I might not be able to easily access their conjunction due to my processing limitations (henceforth also *Quantitative Availability*).

The availability at stake here will be further restricted by features of the social and physical environment: we are supposed to read the newspaper on the table in front of us, but not the letter under the doormat. That's because we can't read everything, and our social environment is such that written testimony is more likely to be present in the newspaper on the table than under the doormat (henceforth also *Environmental Availability*).

I take the availability relation to have to do with a fact being within the easy reach of the knowledge generating cognitive processes of cognizers of the type I belong to. A fact F being such that I am in a position to know it has to do with the capacity of my properly functioning knowledge generating processes to take up F:

Being in a Position to Know: S is in a position to know a fact F if S has a cognitive process with the function of generating knowledge that can (qualitatively, quantitatively, and environmentally) easily uptake F in cognizers of S's type.

Note, crucially, that the account features is an important distinction between being in a position *to know* a fact—where all that needs to happen is that one's cognitive processes pick out the relevant fact—and being in a position *to come to know it*—where more than this is needed: if I need to open my eyes, or turn around, or walk to the other room to pick up the relevant fact, and on the assumption that I can do so, I am in a position to come to know it, but not in a position to know it.

This will help with worries that readers might have that my view makes having evidence too easy, and thereby overgenerates evidence had. It will also help with common worries for broader views than mine, that take potentially knowable facts to be evidence/reasons to believe:[4] On a view like this broader view, the following seems possible. I don't believe p, but I'm in a position to know p (e.g., if I just paid attention to my experience, I'd come to know), I'm in a position to come to know that I know p (e.g., if, having paid attention to my experience I came to know, I'd believe and know that I knew), and I'm in a position to know that I don't know that p (e.g., I know that I don't have an opinion about whether p). If my evidence includes each potentially knowable fact, my evidence includes: p, I know that p, I don't know that p. This, however, is inconsistent and it seems that inconsistent evidence is bad: an inconsistent set of propositions entails everything.

My view escapes this worry, in that it takes evidence to consist of facts that one is in a position to know rather than the broader category of potentially knowable facts, which include facts one is in a position to come to know: if I don't believe p, but I'm in a position to know it, I am not in a position to know that I know p, since it's false: I don't believe it. Compatibly, I may be in a position to come to know it—but that alone does not render p part of my evidence.

But isn't the view too permissive still? Isn't it implausible to think that all the millions of facts that I am in a position to know (think only of the amount of stuff happening within my visual field!!!!) are part of my evidence? No: quantitative limitations on being in a position to know will only deliver *disjunctive* epistemic obligations: Recall: since I can only take up a limited number of facts with my knowledge-generating processes, I will most often be in a situation where I can take up any of F1, F2, F3….Fn, but not their conjunction. In these cases, I will shoulder an epistemic obligation to take up a subset of F1, F2, F3….Fn that is as large as my quantitative take-up limitations. The question that arises is: which is the set that takes normative primacy, and thereby delivers my set of evidence? Availability rankings will deliver the relevant set, on my view: the most easily available subset of facts that I can take up delivers the set of evidence I have: in the case of visual perception, for instance, facts located right in front of me, in the centre of my visual field, which are the brightest, and clearest etc.—in general facts that are most easily available to the cognitive processes of a creature like me.

[4] Many thanks to Clayton Littlejohn for pressing me on this.

Finally: It is easy to see that this view of the having relation delivers the easy answer to what is going wrong in resistance cases: the woman passer-by's testimony in Case 1; the partner's behavioural changes in Case 2; the hands raised by the black students in Case 3, they all constitute facts that are evidential probability enhancers for our subjects and that the subjects in these cases are in a position to know: They are easily available to creatures such as our protagonists, to be taken up by their knowledge-generating processes. These subjects are members of a type of cognizer (human adult cognizer) that hosts cognitive processes with the function of generating knowledge that can easily uptake these facts. Since they fail to do so, their epistemic processes are malfunctioning. This explains the impermissibility intuition. As my grandmother well (hypothetically) put it in the beginning of this paper: what is wrong with these folks is that they have evidence but they fail to update their beliefs accordingly. Finally, it is also clear that this conception of the having relation, in line with the folk notion, will excel at fulfilling its social, moral and political function of predicting accountability: being in a position to know and refusing to take up information will not excuse one from wrongdoing.

3.5 Conclusion

Evidence one has matters, in epistemology and beyond: most centrally, it tells us what information people have available to them when they make their decisions and act, and thereby helps us figure out their accountability status: did they have evidence that what they were doing was wrong? Could they have known that theirs was the wrong decision? If so, they are accountable for what they did. If not, *ceteris paribus*,[5] they may be blameless wrongdoers. The view defended here aims to vindicate this picture, by going against tradition and proposing that we should conceive of evidence had as facts one is in a position to know. The limits of one's skull shouldn't matter for whether one has evidence. Negligent lack of uptake shouldn't give one a free epistemic lunch.

Again, this is, first and foremost, a proposal in conception engineering: as the argument goes, philosophers traffic in the wrong conception of the having relation, and this should change. In a second instance, however, the project may also be one in conceptual engineering: it will depend on one's theoretical commitments (about e.g. the nature of concepts and their permeability to engineering), as well as on whether the current mistaken conception also coincides with us hosting a bad concept of evidence had. If expert accounts have causal power over meanings, expert conception change will help with meaning change.

[5] This clause is important here: one can also be blameworthy for having been an irresponsible inquirer, which resulted in them not having evidence to begin with.

References

Burge, T. (2013). *Cognition through understanding: Self-knowledge, interlocution, reasoning, reflection, philosophical essays*. Oxford University Press.

Comesana, J. (2010). Evidentialist reliabilism. *Noûs, 94*, 571–601.

Derose, K. (2006). "Bamboozled by our own words": Semantic blindness and some arguments against contextualism. *Philosophy and Phenomenological Research, 73*(2), 316–338.

Feldman, R. (2008). Modest deontologism in epistemology. *Synthese, 161*, 339–355.

Kelly, T. (2016). Evidence. In E. N. Zalta (Ed.), *The Stanford encyclopedia of philosophy* (Winter 2016 Edition). https://plato.stanford.edu/archives/win2016/entries/evidence/

Kelp, C. (2018). *Good thinking: A knowledge-first virtue epistemology*. Routledge.

Miracchi, L. (2015). Competence to know. *Philosophical Studies, 172*(1), 29–56.

Pappas, G. (2023). Internalist vs. Externalist conceptions of epistemic justification. In E. N. Zalta (Ed.), *The Stanford encyclopedia of philosophy* (Spring 2023 edition). https://www.plato.stanford.edu/archives/spr2023/entries/justep-intext/

Schellenberg, S. (2018). *The unity of perception: Content, consciousness, evidence*. Oxford University Press.

Simion, M. (2023). Resistance to evidence and the duty to believe. *Philosophy and Phenomenological Research, 108*, 203–216.

Simion, M. (2024). *Resistance to evidence*, Cambridge University Press.

Sosa, E. (2016). *Judgment and agency*. Oxford University Press.

Sylvan, K., & Sosa, E. (2018). The place of reasons in epistemology. In D. Star (Ed.), *The Oxford handbook of reasons and normativity*. Oxford University Press.

Williamson, T. (2000). *Knowledge and its limits*. Oxford University Press.

Chapter 4
Engineering a Concept of Epistemic Justification

Sanford C. Goldberg

Abstract In this paper I argue that the topic of epistemic justification offers promising terrain for conceptual engineering. For one thing, standards of epistemic justification themselves (and the concepts that capture them) aim to address certain practical and theoretical needs, and there is a clear rationale for engineering standards (concepts) to best suit these needs. In addition, neither of the two problems that plague conceptual engineering proposals in other domains—the implementation problem and the change-of-subject problem—seem particularly worrisome in connection with the proposal to engineer epistemological concepts. But the proof will be in the pudding, and in the second half of this paper I propose an engineered concept of epistemic justification. While the concept I characterize here is decidedly non-standard, I argue that it is the one we *should* use, in that it does a better job than rivals at addressing the needs to which a concept of epistemic justification should be responsive.

4.1 1

Conceptual engineering proposals are proposals to improve on the concepts we use. Aiming to address some part of our existing stock of concepts, conceptual engineers typically proceed by enumerating some range of concepts available in a given

With great thanks to Manuel Gustavo, Nate Lauffer, and three very helpful referees for this journal, for extensive comments on earlier drafts.

S. C. Goldberg (✉)
Department of Philosophy, Northwestern University, Evanston, IL, USA

Arché Research Center, University of St Andrews, St Andrews, UK

The African Centre for Epistemology and Philosophy of Science, University of Johannesburg, Johannesburg, South Africa
e-mail: s-goldberg@northwestern.edu

M. G. Isaac et al. (eds.), *New Perspectives on Conceptual Engineering - Volume 2*, Synthese Library 514, https://doi.org/10.1007/978-3-031-98524-9_4

domain of thought, describing the work one or more of these concepts should do as we employ them in thought and speech, and then characterizing the concepts which meet these *desiderata*. If successful, such proposals capture concepts that enable us to speak and judge as we should in the domain in question.

While projects of this sort have become somewhat popular of late,[1] they are not without their critics.[2] Though criticisms are varied, two salient challenges have emerged: the *implementation* problem and the *change-of-subject* problem. According to the former, conceptual engineering is difficult (and perhaps even practically impossible) to implement: whereas the success of a conceptual engineering project depends on subjects actually employing the engineered concepts in speech and thought, conceptual engineers have no special influence over what speakers mean with the words they use in speech or thought, and attempts to affect such change are rarely made in any case. For its part, the change-of-subject problem is based on the worry that, far from improving upon concepts already in use, conceptual engineers simply invent new concepts—in effect, changing the subject.

Both of these objections have received a good deal of attention.[3] I will not be offering a new approach to them. Rather, I want to argue that there is a topic in which these sorts of problems loom less large. The topic in question is epistemic justification.[4] In the first half of this paper I aim to establish, one, that the standard of epistemic justification offers an excellent opportunity to show off the benefits of conceptual engineering, and two, that the results of our labor here are likely to be less susceptible to worries about implementation or change-of-subject allegations than elsewhere. After arguing for this, I will go on in the second half of the paper to engineer a concept of epistemic justification that, I claim, best meets the needs we have for such a concept.

4.2 2

I want to begin by saying a bit about how I will be understanding ‘concept,’ and how this relates to other, nearby notions. By ‘concept’ I mean the type of mental representation used to represent a property in thought. As such it captures the distinctive way that a subject has of thinking about the property in question. Two features are salient here: first, it is a *way of thinking* of a property; and second, it is a way of thinking *of a property*, and the property itself is the concept’s semantic value.

[1] As of March 3, 2023, the *PhilPapers* category for conceptual engineering had 233 entries. (With thanks to a referee for suggesting this measure.)

[2] For recent critiques, see e.g. Deutsch (2020a, b, 2021) and Machery (2021).

[3] For recent work, see e.g. Sawyer (2018, 2020), Belleri (2021), Chalmers (2020), Deutsch (2020a, b), Isaac (2020), Jorem (2021), Machery (2021), Nado (2021a), Nimtz (2021), Pinder (2021), Queloz and Bieber (2022), and Thomasson (2021).

[4] For other conceptual engineering projects in epistemology, see Fassio and McKenna (2015), Haslanger (1999), and Nado (2021b).

Concept individuation, then, is a matter of both of these, so that there can be distinct concepts which share a semantic value.

Both dimensions of concept individuation can be illuminated by contrasting concepts with conceptions on the one hand, and theories on the other.

By 'conception' I mean to include a subject's core beliefs about the property at issue. Such beliefs may well be false. Thus I might think of water as *water*, that is, by way of employing the concept WATER[5] in my thoughts, even as my conception of water includes the belief that water is any clear, tasteless, drinkable liquid that comes out of a tap. We need to distinguish concepts from conceptions to make sure that we get the right semantic value as we characterize what the subject is thinking of.

In addition, I will sometimes speak of *theories* of such-and-such, and when I do I mean a set of sentences aiming to characterize the nature of such-and-such. Thus the theory of justification attempts to characterize the nature of epistemic justification—and so aims to characterize the nature of the concept's target. We need to distinguish concepts from theories to make sure that we capture *the subject's way of thinking* of the target. A correct theory of the nature of an object or property need not be a contribution to the way in which subjects think of the object or property. Again, the case of water is instructive: while the semantic value of WATER is H_2O (or perhaps H_2O in its liquid form), the fact that water = H_2O is no part of the way most people think about the liquid. We might think of a 'way of thinking' of an object, then, as a mental representation which (i) singles out an object in thought (ii) in such a way as to license certain inferences to be made regarding the object. Following the anti-individualist approach to mentality,[6] I do not assume that thinkers are always aware of the inferences that are licensed by the concepts they deploy in thought. On the contrary, for concepts expressed in a natural language like English, speakers may be broadly ignorant of the relevant inferences. What makes it the case that, even so, this concept captures their way of thinking of the object, is that they exhibit deference to some authoritative group of speakers (perhaps the relevant domain experts), and the concept *they* use is correctly so characterized.[7]

Importantly, distinguishing theory from concept is less important when our topic is epistemic justification. Although I will have more to say about this below, a few brief remarks can be made here. Unlike many (most?) other topics, arguably the topic of epistemic justification is not one whose subject-matter is fixed by reality independent of our theorizing about it. On the contrary, 'epistemic justification' is a philosopher's technical term whose usage in epistemic assessment aims to regulate certain practices of ours. We cannot fix the subject-matter at issue without selecting from among the things we expect from an adequate account of epistemic assessment—which is to say, we cannot fix that subject-matter in a way that is fully theory-neutral. For this very reason, contributions to the theory of justification can

[5] I will adopt the practice of designating concepts by employing all capital letters of the word that expresses the concept.

[6] Burge's (1979) defense of anti-individualism is the *locus classicus*.

[7] See Goldberg (2002, 2009) for further articulations of this view.

be construed as proposed contributions to how to think about a certain kind (or certain kinds) of epistemic goodness: such theorizing aims to single out an epistemic property in such a way as to license certain inferences, and to employ the concept just is to think about the kind of epistemic goodness so specified in precisely this way. (I will have more to say on this below.)

4.3 3

There is a clear and compelling rationale for engaging in a conceptual engineering project regarding our concept of epistemic justification. This rationale depends on two plausible assumptions. First, our assessments of epistemic justifiedness aim to address various of our (practical and theoretical) needs. Second, we can evaluate how well a given account of the standards of epistemic justification addresses these needs, with an eye towards determining whether a different account (or various different standards) might do better in this regard. Taken together, these two assumptions suggest a clear rationale for 'engineering' concepts that capture the standards which best address the relevant needs.

Both of these assumptions can be defended.

First, our assessments of epistemic justifiedness aim to address various of our (practical and theoretical) needs. In the spirit of trying to be as inclusive as I can, I submit that the candidate needs which epistemic assessments of justifiedness aim to address include all of the following: determining whether a true belief amounts to knowledge, whether a subject was intellectually responsible in belief-fixation, whether a subject who acted badly or wrongly is to be blamed doing so, whether a subject acted reasonably in the context of the law, whether there is something that the subject should (or should not) believe, and whether proffered testimony was credible. Corresponding to these we have six *desiderata* of a theory of justification[8]: we want a theory which

1. articulates the "core" epistemic condition on knowledge[9];
2. articulates the standards of intellectual responsibility (and/or intellectual duty) in belief-fixation;
3. articulates the epistemic condition on moral responsibility[10];
4. articulates the epistemic condition on (some forms of) legal responsibility[11];

[8] Given what I said at the end of the previous section, these are also *desiderata* on a concept of epistemic justification.

[9] For a defense of the idea that the justification condition is the "core" epistemic condition on knowledge, see Goldberg (2015a, 2018, 2019).

[10] In this connection we can think of the theory of justification as contributing to an account of the conditions on excuses.

[11] "some forms of": those forms of legal responsibility that impose standards of strict liability do not give any weight to the subject's epistemic perspective.

5. addresses "what we should (and/or should not) believe"; and
6. provides standards for determining the credibility of a source on a given occasion of testimony.

To be sure, there will be some theoretical disagreement about which of these should be included, and how they ought to be prioritized. But each responds to a need which epistemic assessment has been thought to address, and each can be seen to inform leading theories of epistemic justification—whether these be axiological or responsibilist, reliabilist, evidentialist, coherentist, dogmatist, virtue-theoretic, or what-have-you.

Second, for each of these *desiderata*, we can evaluate how well a proposed standard of epistemic justification addresses the need in question, and we can compare the various proposed standards with an eye towards determining which does best in this regard. Arguably, this is precisely how the debate about the nature of epistemic justification has evolved. When it comes to articulating the standard of epistemic justification, various criteria have been proposed (see above) and arguments are then made that the proposed criterion best addresses the need(s) in question. Disputes then take the form of debates about the relative weight to attach to the various needs, and which does the best overall job of addressing them.

The points I am making here resemble claims made some time ago by William Alston (1993). Under the rubric of the existence of various 'epistemic *desiderata*,' Alston argued that we should get rid of the concept of justification altogether, and replace it with an account that recognized the broad variety of *desiderata* themselves. For my part, though, I submit that we do better to construe these *desiderata* as stating various candidate *desiderata* we might hope to satisfy in an engineering project. There is of course no assurance in advance that we can satisfy all of these, and if we can't we face the question whether it is better to make do with an engineered concept that captures some but not all of these, or else endorse Alston's suggestion and jettison the concept of justification altogether. But I believe—and will argue in Sect. 4.5 below—that we can capture a good part of what is wanted, and that we do better to employ such a concept than to throw out the concept of epistemic justification altogether.[12]

We can use the two points above to support the claim that the topic of epistemic justification offers promising terrain for conceptual engineering. Various distinct standards of epistemic justification have been proposed. We can think of each such proposal as offering a construal of the subject-matter itself, and so as capturing *a* concept of epistemic justification. And we can evaluate each proposal by how well it addresses (1)–(6) above. Insofar as we find that no single proposal to date addresses all of them adequately, we then face the question whether a better concept (or better concepts) can be engineered. This, it would seem, is a clear rationale for a conceptual engineering project in the theory of justification.

[12] See also Goldberg (2016a).

4.4 4

Not only does epistemic justification offer a topic in which conceptual engineering enjoys a clear rationale, what is more, neither of the two problems that have been raised against conceptual engineering proposals—the implementation problem and the change-of-subject problem—seem particularly worrisome in connection with this topic.

Let us start with the implementation problem. This problem arises insofar as engineered concepts are intended to be employed in speech and thought. This intention can be realized if, and it would seem only if, conceptual engineers can get speakers to employ the concept(s) in question. But that is a very hard thing to accomplish, and in any case conceptual engineers themselves are by no means experts in accomplishing this. We might wonder, then, about the utility of engineering concepts if we can't get people to employ these in their speech and thought.

There are two reasons for thinking that this problem will not be especially acute in connection with epistemic justification.

First, one of the considerations that makes the implementation problem difficult is not present in the context of the concept of justification. In particular, whereas the challenge of implementation is particularly challenging when there is a well-trodden tradition of usage for the terminology in question, when it comes to 'epistemic justification' it is not clear that there is a standard usage in the first place. From the very start this terminology would seem to be something of a philosopher's invention.[13] Here is the way Reynolds (1998, p. 533) makes the point:

> 'Justified', as applied to belief, seems a bit like philosophical jargon, although it is not usually regarded as a technical term. Our intuitions about it are less definite, and so perhaps more open to the influence of philosophical theory, than the intuitions we have about knowledge. I propose to use it as a technical term, explicitly defining it in terms of knowledge. That will allow us to set aside distracting subtleties about its ordinary uses…[14]

Or consider the contrast Tom Kelly makes, in his (2014) Stanford Encyclopedia of Philosophy entry on evidence, between the concept expressed by 'evidence' and that expressed by 'epistemic justification':

> The concept of evidence is central to both epistemology and the philosophy of science. Of course, 'evidence' is hardly a philosopher's term of art: it is not only, or even primarily, philosophers who routinely speak of evidence, but also lawyers and judges, historians and scientists, investigative journalists and reporters, as well as the members of numerous other professions and ordinary folk in the course of everyday life. The concept of evidence would thus seem to be on firmer pre-theoretical ground than various other concepts which enjoy

[13] Empirical evidence for this claim is presented in Grindrod (2022). Phliosophers who have made this point in print include Alston (1985, 1993), Reynolds (1998), Swinburne (2001), Battaly (2001), Kvanvig (2011), Senor (2013), Cohen (2016), Lyons (2016), Eder (2021), and Williamson (forthcoming).

[14] I thank Keith DeRose for bringing this quote to my attention.

> similarly central standing within philosophy. (Contrast, for example, the epistemologist's quasi-technical term 'epistemic justification'.)[15]

What both Reynolds's and Kelly's comments suggest is that there is no standard usage of 'epistemic justification'. Since this is so, there is no standard usage against which one would have to fight as one aimed to implement an engineered concept of epistemic justification. Instead, we might think that there is a term, 'justified', whose core usages are perhaps in ethics and in the law, and whose contours allow for a wide range of possible meanings as applied within epistemology. In short, insofar as there are no established traditions of usage associated with 'epistemic justification' that would need to be resisted in any successful implementation project, the problem of implementation is less severe here than it is elsewhere.

It might be said in response that Reynolds and Kelly can be interpreted in a way that is consistent with the idea that there is a standard usage. Perhaps their point is only that 'epistemic justification' is a term of art, and so at most only a few people participate in the practice of using this term (the epistemologists). This is consistent with there being a standard usage among those who use the term.[16] But in response I would note that if this is so, then the very fact that there are only a few users of the term should make the implementation problem easier to address, since only a few people would need to be convinced of the need to make the change for it to be implemented. Since the people in question are epistemologists—theorists who pride ourselves on following reason where it leads—it seems that this problem is more easily addressable here than elsewhere.[17]

Still, this response does not do away with the implementation problem altogether. Even if there is no established history of usage to create headwinds for the engineer's attempt to implement the new usage of 'epistemic justification', even so, getting speakers to adopt the engineered concept (and to adjust their linguistic usage accordingly) is still hard, and engineers have little or no experience in such attempts. Without such changes, the new concept, even if well-engineered, will remain idle.

This brings me to the second reason for thinking that the implementation problem does not pose a serious a problem in connection with an engineered concept of epistemic justification. Two further points can be made in this connection, both of which reflect the fact that 'epistemic justification' is a term of art.

First, the range of people whose speech and thought would need to change (if the implementation is to succeed) is smaller than in most other cases of conceptual engineering. When it comes to implementing a conceptually engineered concept of epistemic justification, we won't have to convince a large group of speakers, many of whom have no antecedent interest in the matter and many of whom we have no

---

[15] Kelly (2014). Cited 28 December 2022. I thank Chris Tucker for bringing this quote to my attention.

[16] I thank an anonymous referee for this suggestion.

[17] I thank two anonymous referees for pointing out this response to the implementation problem in connection with the concept of epistemic justification.

way of even reaching, to revise how they think and speak.[18] Rather, we need only do so with respect to epistemologists—a well-connected group of specialists who are receptive to good arguments.

Second, when it comes to our concept of justification, the sort of modifications required turn out to be far less substantial than proponents of the implementation worry assume. There are various contexts in which the (or a) concept of epistemic justification has a socially important role to play in epistemic assessment. Even so, the concept of justification is hard to articulate, in ways that will be appreciated by ordinary English speakers. (It is not for nothing that 'epistemic justification' is a philosopher's term of art.) As a result, insofar as ordinary English speakers employ a term that aims to capture a kind of epistemic goodness that is in the vicinity of what is expressed by epistemologists' 'epistemic justification,' often the most charitable way to interpret the speech and thought of such ordinary English speakers will be to construe their usage as expressing what epistemologists express with 'epistemic justification'. What makes this interpretation charitable is precisely that epistemologists' usage of 'epistemic justification' expresses a concept that has been engineered to do the sort of work that we want done in epistemic assessment. Insofar as that engineering work is successful, then, the epistemic engineer has captured what ordinary speakers were trying to say (albeit in a way that was less than fully explicit). What this means, I submit, is that a good part of the problem of implementation is addressed once we have a solid case for thinking that an engineered concept succeeds in capturing the various *desiderata* we should want to capture in this domain. (The remainder of the problem is merely to get epistemologists to start employing the engineered concept.)

Having suggested why I think the implementation problem does not loom quite as large in connection with conceptual engineering projects in epistemology, I move next to consider the change-of-subject problem. I submit that this problem, too, is less threatening of conceptual engineering proposals regarding the concept of epistemic justification. This may be surprising. After all, discussions in the theory of epistemic justification are rife with allegations of changing-the-subject—a familiar allegation so-called epistemic internalists have leveled against epistemic externalists.[19]

Still, I submit that the change-of-subject allegation has less plausibility here than is often supposed. The reason for this has already been anticipated. Allegations of changing the subject require a clear enough sense of the contours of the subject-matter itself, so that one can tell when a proposal has "changed the subject." But when it comes to epistemic justification, there are many candidates for what our term 'epistemic justification' might pick out, and so no single subject matter to respect in this case. (Such a conclusion will likely hold for any technical term where there is considerable long-standing and systematic controversy over how to define

[18] I thank an anonymous referee for making this suggestion to me. I have largely retained their language in spelling this out.

[19] I review the literature, and present a more detailed version of the argument to follow, in Goldberg (2015b).

it.)[20] Insofar as the topic is highly theoretical, all that we have for fixing the subject are the needs which assessments of justifiedness are meant to address. And this means that any serious proposal to capture a standard or concept that aims to address (1)–(6) above can lay claim to being the preferred concept of epistemic justification.

4.5 5

I have been arguing that the theory of epistemic justification offers promising terrain for conceptual engineering. Even so, it remains to be seen whether we can engineer a workable concept of epistemic justification. This is what I aim to do in this section.

As I will understand it here, the question whether we can engineer a workable concept of epistemic justification asks whether we can characterize a concept that can figure in a practice of epistemic assessment, where that practice purports to address the various practical and theoretical needs that give rise to such assessment. (1)–(6) above constitute my attempt to articulate the theoretical tasks that correspond to these needs. While I do not assume that these are the only needs to which our practices of epistemic assessment answer, I do regard them as among the "core" of such needs. In this section, I will venture a proposal that appears to do rather well in connection with (1)–(6)—better, in any case, than standard concepts of justification. Or so I shall argue.

To discern constraints on a workable concept of epistemic justification, it will be helpful to have before us the six theoretical tasks I enumerated above. We expect a workable concept of epistemic justification to be able

1. to articulate the "core" epistemic condition on knowledge;
2. to articulate the standards of intellectual responsibility (and/or intellectual duty) in belief-fixation;
3. to articulate the epistemic condition on moral responsibility;
4. to articulate the epistemic condition on (some forms of) legal responsibility;
5. to articulate "what we should (/should not) believe"; and
6. to provide standards for determining the credibility of a source on a given occasion of testimony.

As I noted above, it is not obvious in advance whether we will be able to frame a single notion that can do all of the above. If we cannot, we face a subsidiary question, whether to make do with the best we have, or to abandon the idea of a single dimension of assessment altogether (à la Alston, 1993; see also Hughes (2021)).

It will be helpful, too, to have a clear sense of the constraints that would need to be satisfied if we are to characterize a concept that manages to do all of (1)–(6). I

[20] With thanks to an anonymous referee for emphasizing this point.

highlight three such constraints. First, the concept will need to be a normative one, which is to say, its characterization will need to establish conditions on *propriety* or *correctness* in connection with all of our epistemic endeavors. Otherwise, it will not be in a position to satisfy either (5) or (6) (and possibly others). Second, the concept will need to be intimately related to the concept of *knowledge*. Otherwise, it will not be able to satisfy (1) or (6) (and possibly others). Third, the concept will have to have a clear connection to the conditions on the relevant sort(s) of *responsibility*. Otherwise, it will not be able to satisfy (2), (3), or (4) (and possibly others). These three constraints are not independent of one another. It seems patent that the conditions of propriety in connection with our epistemic endeavors must be satisfied if one is to count as having attained knowledge. But it is also arguable that there is a kind of intellectual responsibility that is presupposed by the attainment of knowledge as well.

Still, it is important not to overstate the connections between these constraints. It is not obvious whether the sort of responsibility presupposed by knowledge is the sort that figures in (2), (3), and/or (4). On the contrary, corresponding to the various types of responsibility are various demands on the sort of responsibility at issue—and it is unclear at best whether each of these demands is required by knowledge. In bringing this out, I will have occasion to say why it is unlikely that any of the traditional concepts of justification can satisfy all of (1)–(6).

To illustrate, consider intellectual responsibility, the sort involved in (2). If there were a notion of responsibility required by knowledge, presumably this would be it. Yet even here there is a potential disconnect between what is wanted in an account of intellectual responsibility and what is wanted in an account of (the "core" epistemic condition on) knowledge. For example, it might well be intellectually irresponsible to rely on one source of information when another, more reliable source is readily available at little cost. Even so, failure to consult the alternative, more reliable source need not undermine knowledge—and so need not entail that the "core" epistemic condition on knowledge is unsatisfied. Illustrations are easy to come by. Suppose that you calculate the restaurant bill in your head, knowing full well that you are highly competent at mental math and that you are thinking clearly. It would seem that this is a way through which you can come to know how much you owe, even if you fail to avail yourself of the calculator on your cellphone in your pocket.

The same sort of disconnect (between the conditions on responsibility and the "core" epistemic conditions on knowledge) can be seen in connection with other types of responsibility as well. Consider the epistemic dimension of the sort of responsibility that arises in contexts involving allegations of ethical misconduct, the sort involved in (3). Suppose that Sid makes a remark that Nancy finds deeply offensive. When challenged, Sid claims not to have known that his remark would have that effect, and claims further that, while he is sorry for having offended, he had no reason to think that he was deficient in any way in this regard. Nancy responds that Sid's ignorance is precisely the problem, as it exhibits an ethically objectionable lack of care on his part: she expected him to have shown greater concern for the sort of thing that she might find offensive. Now in order to determine whether Nancy's allegation was correct—whether Sid should have shown greater concern—we need to have a sense of what (when it comes to Nancy's sensibility and the likely effects of his comment) he was responsible for having known. But it is far from obvious whether this question can be illuminated by the standards of epistemic justification (as traditionally understood).

It is easy to appreciate the nature of the disconnect here. As traditionally understood, the standards of epistemic justification focus on such things as the subject's total evidence, the reasons on the basis of which he formed and sustained his belief, or (the reliability of) the cognitive processes he employed in belief-fixation.[21] Consider in this light Nancy's allegation that, in assuming that what he said was innocuous, Sid was inattentive to things to which he should have been attentive. Insofar as this allegation is correct, Sid presumably lacks evidence he should have had, and so will be unaware of reasons not to speak (or think) as he did. Now consider the proposal that we should assess Sid's assumption (that what he said was innocuous) by reference to the evidence and reasons in his possession at the time he made his remark. Obviously, such a proposal threatens to beg the very question at issue: in effect, Nancy's allegation is that Sid lacked evidence he should have had, and it is no way to assess this allegation by appealing *only to the evidence he had* in determining whether he should have known better. As in our previous example, so too here it seems that there is a disconnect between the standard of epistemic justification (as traditionally understood), on the one hand, and the epistemic condition on responsibility—in this case, *moral* responsibility—on the other.

At this point those hoping to preserve the connection between the standards of epistemic justification (as traditionally understood) and the epistemic condition on moral responsibility might respond as follows. Either Sid is non-culpably ignorant of the fact that this sort of comment would offend Nancy, or he is *not* non-culpably ignorant of this fact. Suppose he *is* non-culpably ignorant of the fact. Then his ignorance is an excuse, and Sid is not blameworthy for having made the offending remark. Alternatively, suppose that Sid is *not* non-culpably ignorant of the fact.[22] In that case, he is blameworthy for having made the offending remark, as whatever ignorance he has is not an excuse. But—this response continues—in neither case do we need to worry about a disconnect between the standards of epistemic justification (as traditionally understood) and the epistemic condition on moral responsibility. For we can maintain—as part of our proposed analysis of the epistemic condition on moral responsibility—that Sid's ignorance is non-culpable if and only if he lacks the evidence that would make it reasonable for him to suppose that Nancy would find the remark offensive. In this way we see that there need be no disconnect: it is precisely the standards of epistemic justification (as traditionally understood) that diagnose whether his ignorance is culpable or not.

[21] Classic references include Goldman (1979), Pollock and Cruz (1999), and Conee and Feldman (2004).

[22] Either he is ignorant but culpably so, or else he is not ignorant in the first place, and so is aware of the fact in question.

Unfortunately, this response does not save the traditional conception of justification from challenges in connection with (3). To see this, suppose that Sid's total evidence is such that at the time he made the remark he lacked evidence *both* that Nancy would find the remark offensive, *and* that he had duties of care to investigate the matter further. Given his then-current total evidence, he simply had no way of knowing that she would find the remark offensive. Even so, I submit, if the remark in question is the sort whose offensiveness we ordinarily expect people to recognize, we might well regard Sid's ignorance as no excuse *despite* the fact that (given his then-current total evidence) he had no reasons to believe that what he was saying was offensive. We might regard his oblivousness itself as the problem.[23] While this is a controversial view about when ignorance exculpates, it is not implausible to think that we hold one another to objective standards of reasonableness—even if one oneself is not presently in a position to discern all of the considerations that such standards regard as relevant.[24] Consequently, it will not do to assume, as the response above assumes, that the standards of justification, traditionally understood, diagnose when ignorance is culpable.

The conclusion I draw from the foregoing concerns the traditional understanding of justification itself: the concept it yields faces formidable difficulties insofar as it aims to address (2) and (3)—let alone all of (1)–(6). This conclusion paves the way for the prospect of another (non-traditional) concept that can do better in this regard.

There are grounds for hope in this regard. The hope is that we can begin with a traditional concept of epistemic justification and modify it in such a way as to capture a concept which is a candidate for satisfying each of (1)–(6). To this end, let us begin with a standard version of evidentialism,[25] according to which

> EVIDENTIALISM
> For all subjects S, propositions p, and evidence E, where E constitutes S's total evidence at time t, S's belief that p is propositionally justified at t iff E provides adequate support to S's belief that p – that is, iff S's belief "fits" E.[26]

Proponents of EVIDENTIALISM have argued that it does well in connection with (1) (the "core" epistemic condition on knowledge)[27] and (6) (the conditions on testimonial credibility).[28] However, EVIDENTIALISM captures a traditional concept

---

[23] For analyses regarding the conditions on culpable ignorance that allow that one can fail in one's moral responsibility even in the absence of evidence to recognize this, see Harman (2011, 2019). Moody-Adams (1994) suggests similar cases involving "affected ignorance," though it may be that in the sort of case she is interested in, the subjects had the evidence (which is why the ignorance was only "affected").

[24] Compare Scanlon (1998).

[25] I use evidentialism for the purpose of illustration only; if one's preferred traditional theory of justification is different, that might be used as well.

[26] Both EVIDENTIALISM and my own NORMATIVE EVIDENTIALISM (NE) proposed below are concepts of *propositional* justification. They would need to be supplemented with a well-foundedness condition to constitute a concept of *doxastic* justification.

[27] See McCain (2014) and Feldman and Conee (2020).

[28] See Kenyon (2012).

of justification, and I argued above that traditional concepts are not well-positioned to satisfy the responsibilist constraint—that is, they are not well-positioned to satisfy any of (2)–(5). Indeed, this is a familiar charge against EVIDENTIALISM.[29] Still, we might wonder whether EVIDENTIALISM can be modified so as to be better positioned to satisfy the responsibilist constraint.

The engineering task at hand is to engineer a better concept of justification, one that incorporates not only an evidential condition but also a responsibilist condition. To this end, we need an alternative construal of responsibility. Such a construal must be broad enough in scope that it can cover all kinds of responsibility (intellectual; moral; legal). In addition, when it comes to intellectual responsibility in particular, our construal of responsibility should be sufficiently flexible that it can be incorporated into a broadly evidentialist concept of justification.[30] But it cannot be simply a matter of believing in accord with one's own current total evidence, since such a view is susceptible to the worries I raised above in connection with Sid and Nancy. (Sid's problem, recall, was precisely that he lacked the evidence he should have had, and this was not something that was reflected in his own current total evidence.)

I have no argument that there is a *unique* concept of responsibility answering to this description. But what I will argue is that *there is* a concept that does so. According to the concept in question, the conditions on responsibility are determined by *what others would be entitled to expect of us in the domain in question*.[31] Such a conception meets the descriptions above. First, it is broad enough in scope that it can cover all kinds of responsibility: being intellectually responsible is a matter of conforming to the expectations others are entitled to have of one in the intellectual domain, being morally responsible is a matter of conforming to the expectations others are entitled to have of one in the moral domain, and being legally responsible is a matter of conforming to the expectations others are entitled to have of one in the legal domain. Second, this concept of intellectual responsibility can be included as a part of a broadly evidentialist concept of justification. The resulting concept is committed to the following inference: *so long as one has conformed to expectations others are entitled to have of one in the intellectual domain*, one's belief is propositionally justified iff it fits one's evidence.[32] This leaves open how to characterize the conditions on justified belief when the intellectual responsibility condition is *not* met. But we can do so, in a way that continues to treat evidence as the key epistemic notion.

I have tried to defend such a proposal elsewhere.[33] The proposal was this: if in believing that p one fails to conform to expectations others are entitled to have of one in the intellectual domain, then whether one's belief that p is justified is

---

[29] See Kornblith (1983), Cloos (2015), Goldberg (2016b), Miracchi (2019), and Sosa (2021).

[30] For discussions of the nature of intellectual responsibility, and how it might figure in a concept of epistemic justification, see Peels (2016), Goldberg (2017, 2018), and Lackey (2020b, c, 2021b).

[31] The expectations in question are normative, not predictive.

[32] Compare Cloos (2015).

[33] See Goldberg (2017, 2018).

determined (not only by one's current total evidence but also) by the evidence one *would* have had if one *had* so conformed. Call this evidence—the evidence one would have had if one had so conformed—the evidence one "should have had." Then we can say this: if the evidence in E—one's total evidence at the time in question—continues to provide adequate support to one's belief that p even after adding the evidence one should have had, then one's justification is preserved. But if the result of adding to E the evidence one should have had is that E no longer provides adequate support to one's belief that p, then one's justification is defeated. Borrowing a term from Jennifer Lackey[34] I call this sort of defeat "normative defeat," since it reflects defeat by evidence one should have had.

I submit that the foregoing engineers a concept of epistemic justification that does much, and arguably all, that we want from such a concept. The proposal itself (which we might call "Normative Evidentialism," or "NE" for short) is this:

NE

For all subjects S, propositions p, and evidence E, where E constitutes S's total evidence at time t, S's belief that p is propositionally justified at t iff (a) E provides adequate support to S's belief that p *and* (b) E continues to provide adequate support to S's belief that p even when E is supplemented by the evidence S should have had.

(Condition (b) is trivially satisfied if there was no such evidence.) I submit that unlike EVIDENTIALISM, which does well with (1) and (6) but struggles with (2)–(5), NE does well in addressing each of (1)–(6).

Let us begin with (1). NE's first conjunct, (a), is equivalent to the conditions specified in EVIDENTIALISM itself. So assuming that EVIDENTIALISM is not overly-lenient as a characterization of the "core" epistemic condition on knowledge, neither is NE. Still, NE goes beyond EVIDENTIALISM in its second conjunct, (b). But this addition does not make NE too demanding in connection with (1). Consider an agent who fails to have all of the evidence she should have had, where among the evidence she should have had there is evidence that would have rendered her belief unjustified. It should be uncontroversial that such an agent fails to know even if her belief is true and otherwise well-supported by her current evidence[35]; and it is plausible to suppose that the failure to know is attributable here to the agent's evidential situation, which is to say, to the "core" epistemic condition on knowledge. NE seems to get this just right.

Regarding (2), the standards of intellectual responsibility, I submit that NE does well in this regard as well. This should come as no surprise, since NE's condition (b) was proposed precisely in an effort to engineer a conception of epistemic justification that captures considerations of intellectual responsibility. Since condition (b) collects the evidence that a subject would have if she lived up to her intellectual responsibilities, it manages to capture the epistemic significance of intellectual irresponsibility in evidential terms—and thus is broadly evidentialist in spirit.

---

[34] See Lackey (2011, 2020a, 2021a).

[35] For examples illustrating just this point, see the "newspaper case" in Harman (1980) and the unopened letter case in Pollock and Cruz (1989).

It is worthwhile seeing how condition (b) enables NE to address cases of intellectual irresponsibility. To this end, let us consider how NE would handle the mental math/calculator example from Sect. 4.3. According to NE, intellectual responsibility reflects what others are entitled to expect of one in intellectual matters. Accordingly, the nearby availability of a calculator will be relevant to an assessment of a subject S's intellectual responsibility *only if* others were entitled to expect S to have used a calculator in the computation. This verdict seems right: just because there is a nearby more reliable route to truth does not, by itself, settle whether it would be intellectually irresponsible of you not to exploit that route. If you were properly expected to do so, then failure to do so is intellectually irresponsible, but if not, it was not (though it may be described as a failure of epistemic supererogation). Suppose then that it *was* properly expected in the context that the subject S would use a calculator. (Say, S is an accountant and was being paid for her services.) Then, by the lights of NE, we can say that S is answerable not only to the evidence she had in her possession at the time, but also to the potentially defeating force of the evidence she should have had. In such a case S counts as justified if but only if (i) her total evidence supported her judgment and (ii) she would not have gotten contrary evidence had she used the calculator. This battery of justification verdicts seems right.

It is worth underscoring how NE renders the epistemic significance of intellectual irresponsibility. As noted, NE characterizes intellectual responsibility in terms of what others were entitled to expect of one in the intellectual domain, where these expectations are cashed out in terms of the evidence one would have had if one had lived up to those expectations. A failure to live up to these expectations constitutes intellectual irresponsibility; but the bare fact of intellectual irresponsibility, by itself, does not defeat justification. What defeats justification (when it is defeated) is the evidence one should have had. From this we can capture NE's characterization of the epistemic significance of intellectual irresponsibility: such irresponsibility renders one answerable to further evidence, namely, the evidence one should have had.[36] Thus this conception of justification offers us a way to accommodate a responsibilist element into what is otherwise a fully evidentialist concept of justification.

Similarly happy things can be said of NE in connection with (3) and (4).

Regarding (3), the epistemic condition on moral responsibility, NE can plausibly lay claim to capturing this condition. We can appreciate this by extracting from NE an account of the conditions under which ignorance excuses. Given a subject who acted out of ignorance, whether that ignorance excuses (according to NE) will depend not only on her evidence at the time but also on what evidence she should have had. Return to the case of Sid, who acted on the basis of the assumption that his comment was innocuous. At the time he acted on this assumption, he had no reason to think otherwise. But Nancy rightly protested that this was precisely the problem: he should have known otherwise. Insofar as she was entitled to expect as much, then Sid's assumption (that the comment was innocuous) was not justified.

[36] In Goldberg (2017, 2018) I called this sort of context one of "epistemic strict liability."

Since it is that assumption that underwrites his ignorance, his ignorance cannot excuse. More generally, ignorance that reflects an assumption which would be unjustified given the evidence one should have had will not provide an excuse. In this way, NE can allow that whether ignorance excuses depends not only on your current total evidence but also on what others are entitled to expect of you in the intellectual domain. This strikes me as an attractive view of the epistemic condition on moral responsibility.

Moving on to (4), the epistemic condition on (some forms of) legal responsibility, we can reason in the same way. Insofar as "ignorance of the law is no excuse," ignorance of the law will be based on assumptions whose justifiedness does not survive the addition of the evidence one should have had. So, according to NE, this ignorance will not excuse one from legal responsibility—as desired. Still, we might wonder about how NE fares in legal contexts, where a "reasonable person" standard is often employed. Consider a case in which a defendant D aims to repudiate legal responsibility for having injured another person V on the grounds that D was acting in self-defense. If it is to stand up in court, D's claim of self-defense must be reasonable. NE offers us an account of this reasonability condition: a claim is reasonable when (a) it is justified by one's current total evidence and (b) it would remain justified even if we added all of the evidence one should have had. Now, assuming that the "reasonable person" standard is something we are entitled to expect each other to employ, NE's condition will require subjects to have whatever evidence a "reasonable person" would be expected to have. And this, in turn, ensures that no claim of self-defense can be sustained merely because the defendant failed to be aware of something that would be properly expected of a reasonable person. This is a virtue of NE not shared by EVIDENTIALISM, restricted as it is to the evidence the subject has at the time of assessment.

Let us now consider (5), the matter of determining "what we should/should not believe." This is a notoriously vexed topic.[37] To get at the plausibility of NE on this score, we can make clear the range of intuitively correct verdicts NE captures. One relatively uncontroversial claim in this regard is that if S's belief that p isn't justified, then S shouldn't believe that p.[38] If we use NE as our concept of justification, the battery of verdicts yielded in this regard are plausible. According to NE, S's belief that p isn't justified at t if *either* S's belief that p doesn't fit S's total evidence E at t (= condition (a) fails to hold), *or else* there is further evidence S should have had at t which would "normatively defeat" the support E provides for S's belief that p (= condition (b) fails to hold). The first (evidentialist) disjunct, in which (a) fails to hold, captures conditions in which most will agree that S shouldn't believe that p; and the second (responsibilist) disjunct, in which (b) fails to hold, captures

[37] See e.g. Meiland (1980), Heil (1983), Feldman (1988, 2000), DeRose (2000), Chrisman (2008), Goldberg (2017), Booth (2022), Ichikawa (2022).

[38] This isn't entirely uncontroversial; so-called pragmatists will deny it. See Rinard (2019a, b, c) and McCormick (2020).

conditions in which, while it is perhaps more controversial whether S shouldn't believe p, nevertheless a case can be made for thinking that S shouldn't believe that p.[39]

It is here, in connection with the presence of condition (b) in an account of what we should not believe, where we can see the virtues of approaching the concept of justification from a conceptual engineering standpoint. Simply put, the case for including (b) in our characterization of the concept of epistemic justification can be made by making the case for thinking that the resulting concept of justification is one we should employ. In what follows I argue for precisely this.

To begin, when condition (b) fails to hold, we are in a situation in which the subject S failed to live up to the expectations others were entitled to have of her in the intellectual domain, where had she lived up to those expectations S would have acquired further evidence which would have defeated the justification of her belief. Consider the sort of case in which this is likely to happen. These will be cases in which the expectations themselves reflect one's role responsibilities, or responsibilities as a member of a profession or other social group; or alternatively cases where these expectations have emerged in some legitimate social practice in which one is a participant. In such contexts there is a clear sense in which one shouldn't believe as one does: one failed to live up to expectations others were entitled to have of one in the intellectual domain, where had one lived up to them one would have acquired defeating evidence.

Still, some will be unmoved by this.[40] They will insist that it is simply bizarre to suggest the subject *shouldn't believe* what their total evidence supports.[41] This response derives its support from the idea that, when we assess what a subject should believe, we should restrict ourselves to the evidence in the subject's current possession. Clearly, there are contexts in which this restriction makes perfect sense. Consider cases in which, bracketing whether the subject has the evidence she should have, we are exclusively interested in how well she did with the evidence she has. Still, it is important to see that there are all sorts of practical situations in which such a restriction would be inapt, where we have a clear rationale for wanting an assessment that builds in condition (b). Consider contexts like those mentioned above, in which we are relying on an expert, or a professional, or one who occupies a certain social or institutional role, or who is participating in a familiar and legitimate social practice. In these cases, we rely on them precisely because we expect them to live up to the expectations associated with their position or their role. It would be strange to think that such people believe as they should even when they fail to live up to these expectations, under conditions in which had they done so their current total evidence would no longer support their belief. We have a clear rationale, then, for wanting a concept of justification that captures this interest of ours. This is precisely what NE does. I submit, then, that the basis of our preference for NE over something

[39] See e.g. DeRose (2000), Cloos (2015), Goldberg (2017), Miracchi (2019), and Sosa (2021).

[40] What follows is modeled on ideas in Feldman (1988, 2000) and Dougherty (2014).

[41] Juan Comesaña has impressed this point on me, in conversation.

like EVIDENTIALISM itself is one that can be made as part of a proposal to engineer a concept of justification that is more adept at meeting the needs we have in epistemic assessment.

It remains true that we also have a need for the sort of assessment that does restrict itself to a subject's current total evidence. There are all sorts of reasons to want such an assessment. I have no disagreement with this. The only question is whether we have grounds to suppose that such an assessment captures what we should want in a concept of *epistemic justification*. Here I submit that the answer is negative. Again, the perspective of conceptual engineering is helpful. Simply put, (1)–(6) capture the various theoretical tasks corresponding to the practical and theoretical needs to which assessments of justification are answerable. So we should want a concept of justification that satisfies these tasks. I argued above that the concept of justification framed by EVIDENTIALISM fails to do so; and I am currently arguing that the concept framed by NE does do so. If this is correct, then we have grounds for thinking that the concept of justification we should be employing is the one articulated in NE.

So far, in considering NE's ability to capture (5), I have been focusing on verdicts regarding what we should *not* believe. But what does NE have to say about what we *should* believe? The answer is that, by itself, it doesn't say much. However, this is a feature rather than a bug. This is because it is unclear whether we should think that epistemic standards are associated with duties or requirements, as opposed to mere permissions.[42] I submit that the proper answer to this question, whatever it is, can be accommodated by the present proposal. One additional point is worth making in this connection. Whatever we expect from an *affirmative* account of what we should believe, we might well think that any obligations to believe derive, not merely from the epistemic status captured by the concept of epistemic justification, but also from other norms—ethical, professional, social, or what-have-you.[43] If this is correct, then we wouldn't want a concept of epistemic justification, by itself, to do more than underwrite an account of what we should not believe, and be consistent with the correct story about what we should believe. In this respect, NE is in good shape.

Finally, we move to (6), regarding the standards for determining the credibility of proffered testimony. To bring this out we will need to specify a role for the concept of justification in an account of testimonial credibility. A plausible assumption is this: testimony is credible so long as it is proffered on the basis of evidence that warrants justification. In that case, NE entails that a piece of testimony is credible iff (i) it is proffered on the basis of evidence that supports the proposition in

[42] For various recent attempts to defend the idea that there are duties to believe, see Booth (2012), Goldberg (2017, 2022), Basu and Schroeder (2018), Basu (2019, 2021), Lackey (2020b), Crewe and Ichikawa (2021), Ichikawa (2022), and Simion (2025). For those against, see Alston (1988), Feldman (1988), Dretske (2000), Wrenn (2007), and Nelson (2010). See also the collection in McCain and Stapleford (2021).

[43] This is a point that has been made in Goldberg (2017, 2022), McCormick (2020), and Chrisman (2022).

question and (ii) this support is invariant across whatever other evidence the speaker should have had. This is plausible, and as above it seems to outperform EVIDENTIALISM in cases in which we are relying on someone in their capacity as expert or professional, or in their institutional or social role.

And with this I conclude my defense of the idea that NE does well in connection with (1)–(6). The case for regarding NE as capturing the concept of epistemic justification we should employ is not yet clinched, of course; more work would need to be done for that. For example, we would need to show that NE outperforms not only EVIDENTIALISM but also the concepts associated with other leading theories of justification. But I hope to have convinced that NE captures a concept that is worth taking seriously: it has been engineered to be responsive to the various needs we have in employing such a concept.

4.6 6

In this paper I have argued that theory of epistemic justification offers promising terrain for conceptual engineering. In particular, the rationale for engineering a concept of epistemic justification is particularly strong, and the worries traditionally associated with projects of conceptual engineering—the implementation problem and the change-of-subject problem—are less pressing in connection with the concept of epistemic justification. Having made the case for pursuing such a project, I proceeded to offer a candidate concept of epistemic justification. The explicit aim was to engineer a concept that could figure in a practice of epistemic assessment that satisfied six "core" practical and theoretical needs associated with this sort of assessment. One of the big virtues of pursuing conceptual engineering here emerged in my case for the proposed concept. While the concept I characterized is most certainly non-standard in relation to extant concepts of epistemic justification, I suggested reasons to think that this is the concept we should be employing: it better meets the needs in question, and it is no objection to my proposal that it does not correspond to how the term 'justified belief' is used, first, because there is no standard way that the term is used, and second, because this was not the aim in any case.

References

Alston, W. (1985). Concepts of epistemic justification. *The Monist, 68*(1), 57–89.

Alston, W. (1988). The deontological conception of epistemic justification. *Philosophical Perspectives, 2*, 257–299.

Alston, W. (1993). Epistemic desiderata. *Philosophy and Phenomenological Research, 53*(3), 527–551.

Basu, R. (2019). The wrongs of racist beliefs. *Philosophical Studies, 176*(9), 2497–2515.

Basu, R. (2021). Moral encroachment and doxastic wronging. In J. Lackey (Ed.), *Applied epistemology* (pp. 99–118). Oxford University Press.

Basu, R., & Schroeder, M. (2018). Doxastic wronging. In B. Kim & M. McGrath (Eds.), *Pragmatic encroachment in epistemology* (pp. 181–205). Routledge.
Battaly, H. (2001). Thin concepts to the rescue: Thinning the concepts of epistemic justification and intellectual virtue. In A. Fairweather & L. Zagzebski (Eds.), *Virtue epistemology: Essays on epistemic virtue and responsibility* (pp. 98–116). Oxford University Press.
Belleri, D. (2021). Downplaying the change of subject objection to conceptual engineering. *Inquiry*, 1–24.
Booth, A. (2012). All things considered duties to believe. *Synthese, 187*(2), 509–517.
Booth, A. (2022). Ought to believe, simpliciter. *Episteme, 21*(2), 1–13.
Burge, T. (1979). Individualism and the mental. *Midwest Studies in Philosophy, 4*, 73–121.
Chalmers, D. J. (2020). What is conceptual engineering and what should it be? *Inquiry, 63*, 1–18.
Chrisman, M. (2008). Ought to believe. *Journal of Philosophy, 105*(7), 346–370.
Chrisman, M. (2022). *Belief, agency, knowledge*. Oxford University Press.
Cloos, C. (2015). Responsibilist evidentialism. *Philosophical Studies, 172*(11), 2999–3016.
Cohen, S. (2016). Theorizing about the epistemic. *Inquiry, 59*(7–8), 839–857.
Conee, E., & Feldman, R. (2004). *Evidentialism: Essays in epistemology*. Clarendon Press.
Crewe, B., & Ichikawa, J. (2021). Rape culture and epistemology. In J. Lackey (Ed.), *Applied epistemology* (pp. 253–282). Oxford University Press.
DeRose, K. (2000). Ought we to follow our evidence? *Philosophy and Phenomenological Research, 60*(3), 697–706.
Deutsch, M. (2020a). Trivializing conceptual engineering. *Inquiry*, 1–16.
Deutsch, M. (2020b). Speaker's reference, stipulation, and a dilemma for conceptual engineers. *Philosophical Studies, 177*(12), 3935–3957.
Deutsch, M. (2021). Still the same dilemma for conceptual engineers: Reply to Koch. *Philosophical Studies, 178*(11), 3659–3670.
Dougherty, T. (2014). Reassigning responsibilism. In J. Matheson & R. Vitz (Eds.), *The ethics of belief* (pp. 146–168). Oxford University Press.
Dretske, F. (2000). Entitlement: Epistemic rights without epistemic duties? *Philosophy and Phenomenological Research, 60*(3), 591–606.
Eder, A. M. A. (2021). Explicating the concept of epistemic rationality. *Synthese, 199*(1), 4975–5000.
Fassio, D., & McKenna, R. (2015). Revisionary epistemology. *Inquiry, 58*(7–8), 755–779.
Feldman, R. (1988). Epistemic obligations. *Philosophical Perspectives, 2*, 235–256.
Feldman, R. (2000). The ethics of belief. *Philosophy and Phenomenological Research, 60*(3), 667–695.
Feldman, R., & Conee, E. (2020). Evidentialism. In D. Pritchard & R. Neta (Eds.), *Arguing about knowledge* (pp. 174–191). Routledge.
Goldberg, S. (2002). Do anti-individualistic construals of the attitudes capture the agent's conceptions? *Noûs, 36*(4), 597–621.
Goldberg, S. (2009). Experts, semantic and epistemic. *Noûs, 43*(4), 581–598.
Goldberg, S. (2015a). Epistemic entitlement and luck. *Philosophy and Phenomenological Research, 91*(2), 273–302.
Goldberg, S. (2015b). What is the subject-matter of the theory of epistemic justification? In D. Henderson & J. Greco (Eds.), *Epistemic Evaluation: Purposeful Epistemology* (pp. 205–223). Oxford University Press.
Goldberg, S. (2016a). Epistemic justification revisited. *Journal of Philosophical Research, 41*, 1–16.
Goldberg, S. (2016b). On the epistemic significance of evidence you should have had. *Episteme, 13*(4), 449–470.
Goldberg, S. (2017). Should have known. *Synthese, 194*(8), 2863–2894.
Goldberg, S. (2018). *To the best of our knowledge: Social expectations and epistemic normativity*. Oxford University Press.
Goldberg, S. C. (2019). A normative account of epistemic luck. *Philosophical Issues, 29*(1), 97–109.
Goldberg, S. (2022). What is a speaker owed? *Philosophy and Public Affairs, 50*(3), 375–407.

Goldman, A. (1979). What is justified belief? In G. S. Pappas (Ed.), *Justification and knowledge: New studies in epistemology* (pp. 1–25). Reidel.
Grindrod, J. (2022). Justification: Insights from corpora. *Episteme, 21*(3), 794–818.
Harman, G. (1980). Reasoning and evidence one does not possess. *Midwest Studies in Philosophy, 5*, 163–182.
Harman, E. (2011). Does moral ignorance exculpate? *Ratio, 24*(4), 443–468.
Harman, E. (2019). Moral testimony goes only so far. In D. Shoemaker (Ed.), *Oxford studies in agency and responsibility* (Vol. 6, pp. 165–185). Oxford University Press.
Haslanger, S. (1999). What knowledge is and what it ought to be: Feminist values and normative epistemology. *Philosophical Perspectives, 13*, 459–480.
Heil, J. (1983). Believing what one ought. *The Journal of Philosophy, 80*(11), 752–765.
Hughes, N. (2021). Epistemic feedback loops (Or: How not to get evidence). *Philosophy and Phenomenological Research, 106*(2), 368–393.
Ichikawa, J. (2022). You ought to have known: Positive epistemic norms in a knowledge-first framework. *Synthese, 200*(5), 1–23.
Isaac, M. G. (2020). How to conceptually engineer conceptual engineering? *Inquiry*, 1–24.
Jorem, S. (2021). Conceptual engineering and the implementation problem. *Inquiry, 64*(1–2), 186–211.
Kelly, T. (2014). Evidence. In E. N. Zalta (Ed.), *The Stanford encyclopedia of philosophy* (Winter 2016 edition). https://plato.stanford.edu/archives/win2016/entries/evidence/
Kenyon, T. (2012). The informational richness of testimonial contexts. *Philosophical Quarterly, 63*(250), 58–80.
Kornblith, H. (1983). Justified belief and epistemically responsible action. *Philosophical Review, 92*(1), 33–48.
Kvanvig, J. L. (2011). Epistemic justification. In S. Bernecker & D. Pritchard (Eds.), *The Routledge companion to epistemology* (pp. 25–36). Routledge.
Lackey, J. (2011). Acquiring knowledge from others. In A. Goldman & D. Whitcomb (Eds.), *Social epistemology: Essential readings* (pp. 71–91). Oxford University Press.
Lackey, J. (2020a). Punishment and transformation. In E. Lambert & J. Schwenkler (Eds.), *Becoming someone new: Essays on transformative experience, choice, and change* (pp. 230–253). Oxford University Press.
Lackey, J. (2020b). Epistemic duties regarding others. In *Epistemic duties* (pp. 281–295). Routledge.
Lackey, J. (2020c). The duty to object. *Philosophy and Phenomenological Research, 101*(1), 35–60.
Lackey, J. (2021a). Sexual consent and epistemic agency. In J. Lackey (Ed.), *Applied epistemology* (pp. 321–347). Oxford University Press.
Lackey, J. (2021b). *The epistemology of groups*. Oxford University Press.
Lyons, J. (2016). What we talk about when we talk about epistemic justification. *Inquiry, 59*(7–8), 867–888.
Machery, E. (2021). A new challenge to conceptual engineering. *Inquiry*, 1–24.
McCain, K. (2014). *Evidentialism and epistemic justification*. Routledge.
McCain, K., & Stapleford, S. (2021). *Epistemic duties: New arguments, new angles*. Routledge.
McCormick, M. (2020). Believing badly: Doxastic duties are not epistemic duties. In K. McCain & S. Stapleford (Eds.), *Epistemic duties: New arguments, new angles* (pp. 29–43). Routledge.
Meiland, J. (1980). What ought we to believe? Or the ethics of belief revisited. *American Philosophical Quarterly, 17*(1), 15–24.
Miracchi, L. (2019). When evidence isn't enough: Suspension, evidentialism, and knowledge-first virtue epistemology. *Episteme, 16*(4), 413–437.
Moody-Adams, M. (1994). Culture, responsibility, and affected ignorance. *Ethics, 104*(2), 291–309.
Nado, J. (2021a). Conceptual engineering, truth, and efficacy. *Synthese, 198*(7), 1507–1527.
Nado, J. (2021b). Re-engineering knowledge: A case study in pluralist conceptual engineering. *Inquiry, 64*, 76–96.

Nelson, M. T. (2010). We have no positive epistemic duties. *Mind, 119*(473), 83–102.
Nimtz, C. (2021). Engineering concepts by engineering social norms: Solving the implementation challenge. *Inquiry*, 1–28.
Peels, R. (2016). *Responsible belief: A theory in ethics and epistemology*. Oxford University Press.
Pinder, M. (2021). Conceptual engineering, metasemantic externalism and speaker-meaning. *Mind, 130*(517), 141–163.
Pollock, J., & Cruz, J. (1999). *Contemporary theories of knowledge*. Rowman & Littlefield Publishers.
Queloz, M., & Bieber, F. (2022). Conceptual engineering and the politics of implementation. *Pacific Philosophical Quarterly, 103*(3), 670–691.
Reynolds, S. L. (1998). Evaluational illusions and skeptical arguments. *Philosophical and Phenomenological Research, 58*(3), 529–558.
Rinard, S. (2019a). Believing for practical reasons. *Noûs, 53*(4), 763–784.
Rinard, S. (2019b). Equal treatment for belief. *Philosophical Studies, 176*(7), 1923–1950.
Rinard, S. (2019c). No exception for belief. In E. Sosa, J. Fantl, & M. McGrath (Eds.), *Contemporary epistemology: An anthology* (pp. 44–57). Wiley.
Sawyer, S. (2018). The importance of concepts. *Proceedings of the Aristotelian Society, 118*(2), 127–147.
Sawyer, S. (2020). Truth and objectivity in conceptual engineering. *Inquiry, 63*(9–10), 1001–1022.
Scanlon, T. (1998). *On what we owe to each other*. Harvard University Press.
Senor, T. (2013). Justified belief and demon worlds. *Res Philosophica, 90*(2), 203–214.
Simion, M. (2025). *Resistance to evidence*. Cambridge University Press.
Sosa, E. (2021). *Epistemic explanations: A theory of telic normativity, and what it explains*. Oxford University Press.
Swinburne, R. (2001). *Epistemic justification*. Clarendon Press.
Thomasson, A. (2021). Conceptual engineering: When do we need it? How can we do it? *Inquiry*, 1–26.
Williamson, T. (forthcoming). Justifications, excuses, and sceptical scenarios. In F. Dorsch & J. Dutant (Eds.), *The new evil demon*. Oxford University Press.
Wrenn, C. B. (2007). Why there are no epistemic duties. *Dialogue, 46*(1), 115–136.

Chapter 5
Conceptual Ethics and the Categories of "Ideal Theory" and "Non-Ideal Theory" in Political Philosophy: A Proposal for Abandonment

Tristram McPherson and David Plunkett

5.1 Introduction

In contemporary political philosophy, it is common to distinguish "ideal" from "non-ideal" theory, and to explain projects or arguments in terms of this distinction. At the same time, there is a burgeoning literature in which philosophers argue about the relative merits of engaging in "ideal theory" vs. "non-ideal theory". Where one stands with respect to this debate is often treated as a central methodological fault-line in political philosophy and other parts of social and political inquiry. Further, the language of "ideal" and "non-ideal" is infectious, with philosophers now discussing "non-ideal" theories in ethics, epistemology, and the philosophy of language.[1]

This paper makes a prima facie case for *abandoning* use of the terms 'ideal theory' and 'non-ideal theory' in social and political inquiry (across a central range of contexts).[2] Our argument begins by observing two sorts of striking variation. The first is variation in how inquirers characterize "ideal" and "non-ideal" theory. The

[1] See, for example, the papers collected in Tessman (2009). See Cappelen and Dever (2021) for critical discussion of recent work on "ideal" vs. "non-ideal" theory beyond political philosophy, with a focus on the philosophy of language.

[2] In this chapter, we use italics for rhetorical stress, single quotes to mention linguistic expressions, and double quotes for quoting other authors, "scare quoting", and other informal uses (prominently, simultaneous use and mention).

T. McPherson
Ohio State University, Columbus, OH, USA
e-mail: mcpherson.164@osu.edu

D. Plunkett (✉)
Dartmouth College, Hanover, NH, USA
e-mail: david.plunkett@dartmouth.edu

M. G. Isaac et al. (eds.), *New Perspectives on Conceptual Engineering - Volume 2*, Synthese Library 514, https://doi.org/10.1007/978-3-031-98524-9_5

second is variation in the theoretical significance of idealization. Here, there is striking variation across theoretical contexts in the targets that can be usefully idealized, the degree of idealization that can be applied to those targets, and the ways that idealizations can be warranted or worrisome. Against the background of this variety, we argue both that central uses of these terms have the potential to obstruct inquiry, and that there is no theoretically appealing way to reform the existing usage.

It is worth emphasizing that in advocating for abandoning "ideal theory" *talk*, as we do here, we are *not* arguing for a "non-ideal" approach to political inquiry. Rather, as we emphasize in the conclusions, we are arguing that inquirers on all sides of contemporary debates over "ideal theory" would do well to reframe their debates in ways that do not treat "ideal theory" as a central locus of discussion.

We proceed as follows. After more carefully framing our discussion (Sect. 5.2), we present our positive argument, in four stages. First, we introduce two sorts of data: about how the term 'ideal theory' and its cognates are used in contemporary political inquiry (Sect. 5.3) and about the heterogeneous significance of idealizations across theoretical contexts (Sect. 5.4). We then employ this data as the basis for arguing against continuing to use the relevant terms (Sect. 5.5), and against seeking to reform those uses to ameliorate the vices of the existing patterns of use (Sect. 5.6). We conclude by explaining what we take to be the relevant virtues of political inquiry that has abandoned the use of these terms.

5.2 Preliminaries: Conceptual Ethics, Abandonment, and Political Inquiry

This section more carefully introduces our topic, and several crucial assumptions that will guide our discussion. In short, our aim is to evaluate the use of the term 'ideal theory' and its cognates in the context of political inquiry. We begin by briefly introducing how we understand political inquiry. We then introduce the broad sort of project we are engaged in, relative to political inquiry: conceptual ethics. Finally, we introduce a crucial assumption about what sorts of aims we will take to be relevant to evaluating conceptual ethics proposals for terminology within normative political inquiry.

As we understand it, political inquiry is a kind of *normative inquiry*. (Here and elsewhere in this paper, we use 'normative' broadly, to encompass the evaluative, the deontic, and the aretaic, as well as the normative more narrowly construed). On the view that we favor, normative inquiry in general aims to answer extensional and explanatory questions about the instantiation of different normative conditions.[3] When normative inquiry concerning politics is conducted at a high level of generality we are engaged in *systematic* normative political inquiry. For example, when we ask what distributive justice consists in, or what it takes for a state to be

[3] See McPherson and Plunkett (2017, 2021).

legitimate, we are engaged in systematic normative political inquiry. By contrast, when we ask normative political questions tied to certain salient clusters of contexts, we are engaged in *applied* normative political inquiry. For example, if we ask about what rights workers should have in the context of a capitalist society in the twenty-first century, or whether reparations are due to descendants of enslaved people in America given the specific facts of American slavery, we are engaged in applied normative political inquiry. In this paper, we want to evaluate the use of the term 'ideal theory' (and its cognates) in the context of systematic and applied normative political inquiry. To be clear, we take 'non-ideal theory' to be a cognate of 'ideal theory'. So, in making this argument we are as much targeting the use of 'non-ideal theory' as we are targeting the use of 'ideal theory'.

Our argument about 'ideal theory' and its cognates is an instance of *conceptual ethics*. In general, as we understand it, conceptual ethics concerns certain kinds of normative and evaluative questions about our thought and talk, such as questions about which concepts we should use (and why), or which concepts are better or worse (and why). As we understand it, conceptual ethics addresses such questions about concepts, words, or other "representational" or "inferential" devices. Furthermore, it is not limited to the moral or political evaluation of concepts or words, but rather can involve any number of standards for assessing concepts and our use of them (e.g., whether the concepts carve at the joints of reality or not).[4]

Our core question in this paper concerns the evaluation of *language*, not thought. We evaluate the use of the *terms* 'ideal theory' and 'non-ideal theory', *not* the use of the concepts that these terms may be used to communicate. Part of the reason for this focus is that, as we will argue, it is not clear what concepts (if any) these terms pick out.

At a broad level, we can distinguish three competing conclusions that we might draw when engaged in conceptual ethics work on a given term. First, we could conclude that we ought to *retain* the term, with its current meaning and associated patterns of use. Second, we could conclude that we ought to keep the term, but *reform* its meaning and/or use. Or third, we could conclude that we ought to *abandon* use of the relevant term.[5] Our argument in this paper will proceed comparatively, by

---

[4] Our gloss of conceptual ethics draws from Burgess and Plunkett (2013a, b), and Cappelen and Plunkett (2020). On our way of thinking about things, conceptual ethics is closely connected to "conceptual engineering". Put roughly, as we understand it, paradigm instances of work in conceptual engineering draw on or involve work in conceptual ethics, in combination with "conceptual innovation", which involves reforming existing concepts (or words etc.), or creating new ones, as well as work on "conceptual implementation", which involves attempts to get relevant agents to actually take up the proposed conceptual/linguistic changes. This basic take on what conceptual engineering draws from Cappelen and Plunkett (2020) and Burgess and Plunkett (2020). The accounts offered in those two papers differ from each other in some of the details. But those details don't matter here. For some other recent takes on what conceptual engineering involves (and associated discussion of conceptual ethics) see the papers collected in Burgess et al. (2020), as well as Isaac (2020), Pinder (2021), and Chalmers (2020).

[5] Our use of "abandonment" here draws from Cappelen (2023). Note that on some uses of 'eliminativism', to be an "eliminativist about x" is essentially the same as what we are calling "abandon-

arguing that abandonment is preferable to retention or reform, in the case of 'ideal theory' and its cognates.

Proposals to abandon a piece of terminology are often coupled with proposals to replace that terminology with superior terminology. For example, Kevin Scharp's work on 'true' and its cognates advocates for abandoning use of this term in certain theoretical contexts, and instead using the novel technical terms 'ascending truth' and 'descending truth', as appropriate, in these contexts.[6] In this paper, we will not advocate for introducing novel "replacement" vocabulary. Instead, in the conclusion we will suggest that we can use other existing vocabulary to clearly focus attention on the virtues and vices of particular candidate idealizations in particular theoretical contexts.

A central question for any conceptual ethics project is: what normative or evaluative standards should we use in assessing the use of the target words or concepts? This is, in general, a deep and difficult question.[7] Here we tentatively propose an approach to answering it, in our context. This approach begins by assuming that one important way to evaluate certain words and concepts, as they are used within a particular branch of inquiry, is by reference to the success conditions (or "aims") of that inquiry.

For simplicity's sake, we will make the further assumption that the success conditions of both "systematic" and "applied" political inquiry are *alethic* or *epistemic*. That is, a normative political inquiry is successful when, and to the extent that, it results in things like true belief, knowledge, or understanding concerning its focal topic(s). Thus, when evaluating our target terms, we will be interested in how well their use serves the project of coming to have true beliefs, knowledge, or understanding of the relevant normative topics.

This approach is admittedly controversial: some political inquirers appear to think that at least some political inquiry has directly *practical* aims rather than alethic or epistemic ones.[8] For example, perhaps one might hope that political inquiry could help to guide our political lives together, or to promote justice—and that one of these goals might itself set the success-conditions for the inquiry as such.[9] We have focused on broadly epistemic aims for three reasons. First, we think

ing" either the term 'x' or a concept tied to our use of 'x'. However, in other contexts, 'eliminativism about X' denotes a nihilist thesis in metaphysics: roughly, the denial that Xs exist. It is important to distinguish the nihilist metaphysical thesis from the conceptual ethics abandonment thesis, as emphasized in Burgess and Plunkett (2013b). For ease of exposition in this paper, we stick with the terminology of "abandonment" rather than "eliminativism".

[6] See Scharp (2013).

[7] For discussion, see Burgess and Plunkett (2013b) and McPherson and Plunkett (2024a).

[8] Consider another non-epistemic role for usage of this terminology: saying "I am a [non-] ideal theorist" might function to signal one's allegiance to a broad camp within the academic community, in the same way that saying "I am an *analytic* philosopher" might. We think this sort of signaling usage is sociologically important. However, it is not our focus in this paper, and we thus set it aside in what follows.

[9] For some versions of this idea, see Haslanger (2000), Geuss (1981), and Jaggar and Tobin (2013). For discussion, see Plunkett (2016).

these aims are plausibly important to a wide range of political inquirers. To underscore this point, one might well think that engaging in political inquiry (where it is understood to have purely epistemic success conditions) could be instrumentally helpful for achieving other, practical goals (such as the promotion of a more just society), even if those practical goals aren't part of the aims of inquiry as such. Second, it's not feasible, given the length of this paper, to develop our argument relative to the full range of possible aims for political inquiry. And, finally, we are optimistic that the conclusions that we draw could be generalized to apply to other conceptions of political inquiry, on which it has non-epistemic aims.

We take it that epistemic or alethic aims can provide a plausible and entirely familiar rationale for the abandonment of words or concepts, especially for theoretical words or concepts. Consider, for example, terms like 'phlogiston' (in chemistry) or 'hysteria' (in clinical psychology). In both cases, abandonment seems decisively warranted, due to the fact that use of these terms has turned out to be counterproductive for investigating the relevant parts of reality.[10]

With this take on the aims of inquiry in hand, we can sharpen our central conceptual ethics question about the term 'ideal theory' and its cognates, in contemporary political inquiry. Given our epistemic aims, is it best to retain these terms with their current meanings, to reform the use and meanings currently associated with these terms, or to simply abandon their use? This paper argues that contemporary political inquirers would (in general) be better off relative to epistemic aims if they abandon use of the term 'ideal theory' and its cognates. Note that this thesis is a generic claim, not a universal claim. Like most plausible abandonment theses, it is compatible with the possibility that there are some (unusual) contexts where it would be epistemically useful for those engaged in political inquiry to continue to use this terminology. Furthermore, it is worth underscoring that since our focus here is on abandoning the use the terminology of 'ideal theory' and its cognates in the context of engaging in political inquiry, our main argument is compatible with not abandoning that terminology in other contexts (e.g., perhaps in certain contexts when engaged in intellectual history).[11]

With these clarifications in hand, the next two sections prepare our case for abandonment by providing two sorts of relevant data: data about the heterogeneity of the contemporary *use* of 'ideal theory' and its cognates, and data about the heterogeneous *functions* that idealizations can play in political inquiry.

[10] Note that the case of 'hysteria' helps to illustrate that this sort of epistemic standard is not the only standard relevant for conceptual ethics. Another excellent reason to abandon 'hysteria' was that use of this term functioned to reinforce sexist norms and practices.

[11] We suspect that, if our main argument in this paper is right, then, for many contexts in discussing intellectual history, it will be better to mention (rather than use) the terminology of 'ideal theory' and its cognates. But we don't take a stand on that in this paper.

5.3 The Heterogeneous Use Data

'Ideal theory' and its cognates were arguably introduced as technical terms in political inquiry by John Rawls's *A Theory of Justice*. For Rawls, engaging in "ideal theory" about justice involves asking what just social/political institutions would be like given (among other things) certain *idealizing assumptions*.[12] Rawls appears to make two of these assumptions definitional of "ideal theory". The first is the assumption of full compliance with just institutions by the agents who are regulated by those institutions.[13] The second is that we are to consider a society in circumstances favorable to the establishment of justice.[14] Rawls also makes other idealizing assumptions in *A Theory of Justice*, although he does not appear to make them definitional of "ideal theory". For example, he assumes that agents in the society he describes are not motivated by envy in the way actual humans often are.[15]

Given that 'ideal theory' and its cognates were introduced as technical terms, it is a striking fact that contemporary political inquiry has not tended to simply defer to Rawls' characterizations. Instead, the contemporary landscape is rife with seemingly cross-cutting characterizations. Consider representative framings from three recent surveys of this literature:

- "Recent debates have seen the term used in different ways and its critics attacking a variety of different targets."[16]
- "A quick glance at what falls under the heading 'ideal/non-ideal theory', however, reveals the heterogeneity of this debate."[17]
- "The notion of an ideal theory can have at least two different meanings.... First, a political theory can be ideal in that it provides an ideal, or goal that we should aim for.... Second, a political theory can be ideal in that it involves models intended to represent certain phenomena, with the associated false assumptions (either abstractions or idealizations)."[18]

To get a feel for the heterogeneity of the use of these terms, consider Laura Valentini's influential attempt at regimentation. Valentini suggests that there are "at least three distinct meanings given to the adjectives 'ideal' and 'non-ideal'" in this

[12] We do not also offer a gloss on Rawls' characterization of 'non-ideal' theory, because we take the text here to be less clear. Simmons (2010) reads Rawls as offering a comparatively narrow account of "non-ideal theory", where, roughly, it concerns the steps we can take on the path towards ideal justice. Some passages in Rawls, however, suggest a broader characterization of "non-ideal theory", including Rawls' claim that such theory concerns "which principles to adopt under less happy circumstances" (Rawls, 1971/1999, p. 216).

[13] Rawls (1971/1999, e.g. at 8; p. 216).

[14] Rawls (1971/1999, p. 216).

[15] Rawls (1971/1999, §25).

[16] Stemplowska and Swift (2012, p. 373).

[17] Valentini (2012, p. 654).

[18] Thomson (2020, p. 2).

debate.[19] The first is the Rawls-inspired idea that ideal theory involves an assumption of *full compliance*. The second is the idea that an "ideal theory" is "utopian" rather than "realistic". This means, roughly, that it is insensitive to hypotheses about feasibility. Finally, there is the idea that an "ideal theory" gives us an "end state" or goal to work towards, whereas "non-ideal theory" is a theory of how to transition towards that goal.

Even Valentini's expansive regimentation, however, fails to capture the full range of prominent characterizations in the literature. Consider some examples.

First, Zofia Stemplowska offers a characterization of the distinction between "ideal" and "non-ideal" theory as concerning whether a theory offers *real-world guidance*. She argues that "one helpful way of understanding the cut between ideal and nonideal theory is roughly to classify as nonideal any theory that issues recommendations that are desirable and achievable for us given where we find ourselves and, as ideal theory, any other type of normative theory."[20] At first blush, Stemplowska's distinction does not fall neatly into any of Valentini's broad categories. For example, a "realistic" theory could refrain from offering guidance, while one could offer guidance based on a "utopian" theory. And a familiar thought in the contemporary philosophical literature is that one could offer practical guidance that is not directed towards getting us closer to some "ideal" end state.

Second, consider Elizabeth Anderson's characterization of the distinction. Anderson writes: "In ideal theory, ideals.... are not subject to testing in practice because they set standards, outside of practice, for the success of practice."[21] Anderson claims that non-ideal theory contrasts with this as follows: "In nonideal theory, ideals embody imagined solutions to identified problems in a society. They function as hypotheses, to be tested in experience."[22] Anderson's distinction concerns the *epistemological status* of ideals relative to practice and experience. It thus seems to crosscut both Valentini's broad categories and Stemplowska's distinction.

Finally, prominent discussions disagree about whether it is appropriate to classify certain debates as "ideal theory" debates. Consider two significant examples. First, some classify G. A. Cohen's insistence that political inquiry should bottom out in "fact-insensitive" normative principles as a kind of "ideal theory".[23] Others, however, take these claims from Cohen to be orthogonal to the "ideal/non-ideal" cut.[24] Second, while (as we have seen) Valentini treats sensitivity to feasibility as

[19] Valentini (2012, p. 654).

[20] Stemplowska (2008, p. 339).

[21] Anderson (2010, p. 6).

[22] Anderson (2010, p. 6).

[23] For example, see Valentini (2012, §2.1) and Thomson (2020, p. 2), discussing (Cohen, 2003, 2008).

[24] For example, see Enoch (2018, 11n.11). It is also natural to read (Stemplowska & Swift, 2012, §4) as endorsing the view that Cohen's fact-insensitivity thesis crosscuts the "ideal/non-ideal" distinction.

one of her three central characterizations of "the" "ideal/non-ideal" cut, Christopher Thompson suggests that this is a related but distinct issue.[25]

A further point is worth emphasizing here: on many (though not all) prominent ways of drawing the distinction between "ideal" and "non-ideal" theory, "non-ideal theory" marks out something more specific than just the negation of "ideal theory". For example, consider Anderson's epistemological way of drawing the distinction. On her view, "non-ideal theory" involves a *specific* kind of epistemological alternative to what she takes to be the one involved in "ideal theory", rather than encompassing the totality of possible alternatives. Or, to take another example, consider Tommie Shelby's gloss of the distinction between "ideal" and "non-ideal" theory. In short, he claims that a) ideal theory aims to provide "a comprehensive account of the principles a society must satisfy to be fully just"[26] and b) non-ideal theory "specifies and justifies the principles that should guide our responses to injustices".[27] The latter isn't just *any* kind of theory that isn't the former. Rather, it's a specific kind of other theory, which, on Shelby's view, is meant to complement the former in an overall theory of justice.[28]

This survey is far from exhaustive.[29] But it allows us to draw a striking conclusion. Careful observers are broadly united in concluding that there are multiple substantially distinct prominent uses of 'ideal theory' and its cognates in recent political inquiry. However, they also disagree about (i) how to regiment that range of uses, and (ii) which sorts of views even fall within the scope of discussions of "ideal vs. non-ideal theories". The diversity is significant enough to these commentators that, as we have seen, two of the three recent surveys of this literature we discuss above (Valentini and Thompson) seemingly suggest that the relevant terms are *polysemous*, meaning that they have multiple circulating meanings. The authors of the third (Stemplowska and Swift) frame their discussion by describing the recent literature as marked by "confusion", "ambiguities", "talking past", and "cross-cutting complexities".[30] David Enoch sums up the state of play as follows: "The terms "ideal theory" and "non-ideal theory" have unhelpfully come to stand for several different things".[31]

---

[25] For example, see Thomson (2020, p. 12).

[26] Shelby (2016, p. 11).

[27] Shelby (2016, p. 11).

[28] As he puts it: "Ideal theory and nonideal theory are complementary components of an endeavor to devise a systematic account of social justice" (Shelby, 2016, p. 11).

[29] In addition to not canvassing every view about how to distinguish "ideal" from "non-ideal" theory, we haven't introduced other important aspects of the debate, such as discussion of the primary *purpose(s)* of each kind of theory. For some recent discussion here, see Shelby (2016), McKean (2017), and Stemplowska (2017).

[30] Stemplowska and Swift (2012, pp. 373–374).

[31] Enoch (2018, p. 11).

5.4 The Heterogeneous Function Data

We have just introduced the remarkable heterogeneity of the ways the term 'ideal theory' and its cognates are used. This heterogeneity is compatible with the hypothesis that there is a singular way we *could* use this term, which would be especially theoretically valuable. This section introduces reasons to doubt this hypothesis. We suggest that there is not a single determinate thing that it would be useful for 'ideal theory' to pick out. We make our case by focusing on the central notion of idealization. We sketch four relevant sorts of theoretically important variation with respect to idealization in political inquiry.[32] Such idealizations vary with respect to:

- what is idealized;
- the degree of idealization;
- the value served by idealization; and
- the potential liabilities of idealization

We will consider these dimensions in turn.

First, as we have briefly illustrated in the preceding section, political inquirers can and do idealize in a wide variety of ways. For example, they sometimes assume *full compliance* with political norms, or that political agents are *rational*, or *self-interested*, or *informed about politically relevant facts* or *not prone to envy*, etc. They assume that the polity does not have a politically non-ideal *history* with which it needs to wrestle, and that it does not exist in a world that poses an *existential threat* to polities with certain political structures.[33] Further, these *actual* idealizing assumptions are only the tip of an iceberg of *potential* idealizing assumptions that it might be useful to make in political inquiry. Call this point the *variety of dimensions of possible idealization*.

Second, with respect to any dimension of idealization, a theory can build in a more or less demanding idealizing assumptions. For example: consider informedness. We might idealize modestly, assuming that citizens are all (e.g.) informed about the constitutional basics of their polity. Or we might idealize more boldly, assuming that the polity is *free of misinformation*, or, more boldly still, that citizens are *omniscient*. Similar points apply to other dimensions of idealization. For example, it might be illuminating for some purposes to theorize about certain political questions under assumptions of perfect compliance, but, for other purposes, it might be illuminating to engage in political inquiry that assumes a non-actual but feasibly achievable level of compliance.[34] More generally, a theory can be more or less

[32] O'Neill (1996, Chap. 2) helpfully distinguishes *idealization* (which, on her account, involves theorizing using claims that are not true) from *abstraction* (which, on her account, involves ignoring certain true claims). We take much of what we say in this section to be adaptable, mutatis mutandis, to abstraction.

[33] Versions of each of these assumptions are made in Rawls (1971/1999) and much of the literature following and engaging with that text.

[34] In connection with this point, it is instructive to consider the rule consequentialism literature, where there is a vigorous debate concerning whether the relevant rules should be tested relative to

"realistic" or "utopian" in its assumptions about human motivation, the material conditions we face, etc.

Third, idealizations can be valuable for inquiry for different reasons. Consider three examples.

(a) An idealization might be *constitutive* of the normative category that we are investigating. For example, in *The Moral Problem,* Michael Smith argues that facts about what certain kinds of idealized agents (ones with more information, who do not make certain kinds of reasoning errors) would want our current "non-ideal" selves to do *constitute* facts about what normative reasons for action we have.[35] In political inquiry, certain idealizations might similarly be argued to be constitutive of a perfectly just regime.
(b) An idealization might be *epistemically valuable for inquiry*, by helping us to more efficiently arrive at relevant knowledge, understanding, or some other epistemic goal. For example, if we are wondering about how to evaluate the actions of a particular university's administration, it might be helpful to abstract away from certain aspects of the university's politics, in order to get a deeper critical perspective on the overall priorities of the university. Similarly, some have argued that understanding the "ideally just" is important for evaluating practical proposals for political improvement.[36]
(c) An idealization might be *explanatorily illuminating* in some further way. Suppose there are explanatory relations that do not simply reduce to epistemic or constitutive relations. If so, then idealizing facts could play one of these roles. (For example, on some metanormative views, certain naturalistic facts can *ground* normative facts, without partially *constituting* them.)[37] Idealizing facts might well play such a grounding role. For example, one might think that *what grounds* one (imperfectly just) state being preferable to another is their relation to certain facts about ideally just states.

Fourth, idealizations can sometimes be theoretically *unhelpful* or *counterproductive* in political inquiry, for several reasons. Consider three:

(a) An idealization might simply fail to help answer the question in political inquiry that most interests us. For example, Amartya Sen argues that, in order to evaluate real-world political outcomes and policies, we need to be able to evaluate the *comparative* justice of alternative outcomes we could achieve. He uses an analogy to argue that perfect justice is often simply irrelevant to this task: if I want to know which of two mountains is taller, knowing that some third mountain is the tallest in the world is orthogonal to my task. In a similar way, knowing

(e.g.) *compliance* or *acceptance*, and what degree of compliance (etc.) is relevant. For discussion, see Hooker (2023, §6).

[35] Smith (1994).

[36] Simmons (2010, pp. 34–36).

[37] For example, see Berker (2019).

that some state C is perfectly just may be totally unhelpful to knowing which of A or B is *more just*.[38]

(b) Straightforward inference from normative premises about idealized circumstances to normative conclusions about realistic ones can be unreliable, in light of the "problem of the second best".[39] To adapt David Enoch's example, speed limits that would be optimal under full compliance with traffic laws might be disastrous given how people are actually disposed to drive.[40]

(c) Some idealizations eliminate the very features that partially constitute the most pressing political challenges that we face. This elimination, it has been argued, can play an ideological function.[41] As such, focusing our attention on debates about the norms that regulate such an ideal society can make it possible to engage systematically in normative political inquiry without ever addressing the pressing question of what justice requires *of us*, in the unjust here and now.

Together, the first two dimensions of variation suggest that there is an exceptionally broad range of possible degrees of idealization, with various target dimensions, that might appear relevant to political inquiry. The third and fourth dimensions of variation suggest that different nodes across this range will vary strikingly in whether and how they are valuable, useless, or counterproductive.

To bring this point out, consider two examples. First, *some* normative political questions are constitutively tied to idealizations, while others are not. Contrast, for example, the question of what is compatible with *perfect justice* with the question of whether a certain decision made by the Supreme Court in Canada is *more or less* just than an alternative decision it might have made. Even if Sen is correct that certain idealizations are irrelevant to answering the second question, they might be indispensable for answering the first question. Second, suppose that achieving meaningful reparations for slavery is in fact "politically infeasible" in the current social/historical circumstances in the USA. If true, this reality might imply that pushing for reparations is not a useful proximate goal for political action. But nonetheless, idealizing away from these feasibility facts might play a liberatory, as opposed to oppressive, role in our political thought.

---

[38] Sen (2006).

[39] The foundational text for discussion about "the problem of the second best", in the context of theorizing pareto optimality, is Lipsey and Lancaster (1956–1957). For recent discussion about the importance of this problem for normative work in political philosophy, and for further references to other recent discussions, see Wiens (2020).

[40] See Enoch (2018, p. 12).

[41] Mills (2005).

5.5 The Case Against Retention

In this section, we draw on the heterogeneous use data and the heterogeneous function data to argue against retention of the term 'ideal theory' and its cognates in political inquiry.

Before we begin, we want to note some general reasons in favor of retaining a piece of technical terminology, which apply to the current use of 'ideal theory'. These include benefits that flow from familiarity with the terminology and concepts (and what they refer to) within the relevant community of inquiry, and ones that flow from smoother engagement with the existing literature that employs that terminology and those concepts.

We want to emphasize that these benefits set a relatively low bar. For example, shortly after a more accurate theory of combustion was developed, use of 'phlogiston' retained these virtues. But, these virtues were clearly outweighed by other considerations, which favored abandoning use of the term 'phlogiston' for the purposes of scientific inquiry. These considerations included, crucially, that the new theory no longer posited the existence of phlogiston, and that people had no good independent evidence in favor of believing in phlogiston. We don't take the case for abandoning 'ideal theory' and its cognates for the purposes of political inquiry to be as strong as the case was for abandoning 'phlogiston' for the purposes of scientific inquiry. However, we argue that, as in the 'phlogiston' case, there are considerations that outweigh the modest considerations we just canvassed in favor of retention, and point in favor of abandonment.

Our argument draws on three central ideas: the standards for evaluating technical terminology, the unhelpfulness of the pattern of usage characterized in Sect. 5.3 above, and unreliable inferences that this existing pattern can lead to.

Our first line of argument appeals to the fact that 'ideal theory' and its cognates are *technical terms*, like 'supervenience' but unlike, say, 'justice'. We think this fact is relevant to how we evaluate whether to retain a piece of terminology in at least two ways.

The first issue is that there are dangers associated with abandoning folk terminology that do not apply to abandoning technical terminology. To illustrate this point, imagine for a moment a proposal that political inquirers abandon the use of 'justice'. There are several deep worries one might have about this proposal, in light of the fact that 'justice' is a central folk term for thinking about politics. For example, ceasing to use 'justice' would risk decoupling political inquiry from an important topic that ordinary people care deeply about with respect to social/political life.[42] Further, thinking in terms of 'justice' might be *epistemically crucial* for many people: it might be that much of our competence in thinking about normative political

---

[42] For similar worries, see Jackson (1998), as well as Strawson (1963)'s famous concerns about Carnapian explications (of the sort discussed in Carnap (1947/1956)). For connected discussion, see McPherson and Plunkett (2024b).

questions is psychologically encoded in terms of 'justice'. These worries do not seem to apply to abandoning a technical term.

The second way in which the fact that 'ideal theory' and its cognates are technical terms is relevant is that we characteristically introduce and retain technical terms in order to achieve certain broadly epistemic goals such as (1) the ability to track joint-carving distinctions and (2) increased clarity, precision, and usability.

Consider the first goal: sometimes it can be valuable to introduce and retain a technical term because doing so helps to carve the relevant domain at its joints, which we have epistemic reason to want to do.[43] Presumably, we successfully and usefully introduced 'atom' in this way, despite our having striking false beliefs about the nature of atoms for centuries after its introduction. One upshot of Sect. 5.4, however, is that there is no reason to believe that there is a single theoretical joint in the vicinity of contemporary uses of 'ideal theory'. Rather, what we see is a wide range of possible idealizations that could potentially be theoretically useful to track in particular contexts.

Now consider the second goal. Often, useful technical terms or explications have meanings that are *more precise and explicit* than alternative terminology, or are *more usable* (e.g. in having a clear operationalization that can be used for certain inquiries). But as we emphasized in Sect. 5.3, what we see in the case of 'ideal theory' and its cognates is instead a forest of cross-cutting distinctions. We conclude from this that contemporary use of 'ideal theory' lacks the sorts of virtues that typically support retaining a technical term.

The second central strand of our case for abandonment appeals to the pattern of contemporary usage summarized in Sect. 5.3. The simplest case for abandonment here is that this existing pattern is, frankly, a mess. But we can say more.

One possible consequence of the heterogeneity of usage is that the term 'ideal theory' and its cognates simply fail to refer. One way to motivate this idea is as follows: if the meaning of our words depends in some way on how relevant "experts" use words, and there is massive divergence in the expert usage of a given word, then that might yield the result that the word fails to refer.[44] If the term 'ideal theory' fails to refer, that would constitute a powerful and familiar reason for abandoning the term.[45]

---

[43] It should be noted that, on some ways of thinking about various possible values in conceptual ethics, a concept's being "joint-carving" is a primarily "metaphysical", rather than "epistemological", good. See, for example, the discussion in Burgess and Plunkett (2013b). We don't want to wade into debates here about how to best characterize these different goods. Our point here is just that, in many contexts, we might well have epistemic reasons to track an important metaphysical distinction.

[44] For discussion of this sort of possibility, including an argument that it might apply to many more words than one might initially think, see Cappelen (2013).

[45] This interacts with the fact that 'ideal theory' and its cognates are technical terms. Some philosophers argue for retaining certain folk terms that they believe fail to refer because of the usefulness of other aspects of their widespread use. For an example of this, in the case of moral terminology, see Olson (2014).

Another credible possibility is that 'ideal theory' manages to refer, but is *polysemous*: that is, the term as ordinarily used has a range of different meanings (this idea is seemingly suggested by two of the survey articles we discussed in Sect. 5.3). Polysemy is plausibly a striking vice in a technical term, especially if (as is the case here) the relevant meanings are quite different from each other.

Perhaps the most promising hypothesis for the retentionist is that, on current usage, 'ideal theory' and its cognates are *context-sensitive* expressions, with facts about the context of use fixing which of the extremely heterogeneous collection of apparent meanings for the term is expressed in that context. It is unclear how promising this is as a hypothesis about the actual semantics of 'ideal theory'. For our purposes, we will simply grant this hypothesis for the sake of argument and ask: if the hypothesis is true, does that give us reason to retain the term?

In our view, it is generally undesirable for technical terms to be context-sensitive, for several reasons. For example, a context-sensitive term will not consistently carve nature at the joints. And it will not provide precision and clarity in a transparent way. In light of these points, we think there has to be a specific rationale for embracing a context-sensitive technical term. It may be that it is sometimes fruitful to have technical terms that allow inquirers to coordinate on different intensions in different contexts.[46] But we see no such rationale in the context of 'ideal theory'. To foreshadow our constructive alternative: there is no evident practical need in political inquiry to coordinate in this way, as opposed to simply focusing on discussing and evaluating *specific* idealizations.

The final strand of our argument against retaining 'ideal theory' and its cognates begins with the idea that the heterogeneity of the existing pattern of usage associated with 'ideal theory' may promote various sorts of unreliable inferences.[47] The basic concern here is as follows. The heterogeneous use data suggest that a wide variety of contrasting uses of 'ideal theory' are salient in contemporary political inquiry. This variation raises a danger that inquirers could make unreliable inferences, by drawing conclusions based upon unintended but psychologically salient uses.

Consider one potential example of this sort of unreliable inference. When making idealizations, a normative theorist should be careful that these idealizations—and the rationales for deploying them—fit together in a coherent way. For example, one shouldn't be assuming some things that are wildly infeasible, while insisting on high standards of feasibility for other dimensions—at least without a good explanation for this discrepancy.[48] Framing one's discussion in terms of "ideal theory",

[46] For an argument for this kind of view about the term 'topic continuity' in conceptual engineering (and beyond), see McPherson and Plunkett (2024b).

[47] The worry we describe below about unreliable inferences connected to uses of 'ideal' and 'non-ideal' theory is an instance of the kinds of general worries about unreliable inference that we discuss in McPherson and Plunkett (2020).

[48] For a case study involving how (often subtle) switches in the kinds of idealization used in making arguments in normative political philosophy can matter to assessing those arguments, see Plunkett (2024), discussing (Rose, 2020).

rather than the particular idealizations at issue, can prevent us from carefully tracking how well the idealizing assumptions one is deploying fit together.

Consider a different way that 'ideal theory' may promote unreliable inference. The various heterogeneities discussed in Sects. 5.3 and 5.4 strongly suggest that there is not a single, theoretically interesting topic in the vicinity of existing discussions of "ideal theory". But the existence of a technical term can encourage the illusion that there is such a topic. One upshot of this illusion is that various authors sometimes disagree about whether certain topics are part of "the" "ideal theory/non-ideal theory" debate. In Sect. 5.3, we mentioned two examples of this: disagreement over the classification of Cohen's fact-insensitivity thesis, and disagreement over the classification of theories that are sensitive to issues of feasibility. The problem is this. We have suggested that there is no underlying theoretical unity in existing talk of "ideal theory". But in the absence of such unity, it is unclear what compelling theoretical basis there could be for insisting that there is some mistake being made in either including or excluding these topics from the topic of "ideal theory".

One further reason that unreliable inference is a substantial danger is that the considerations that might justify one kind of idealization (say, assuming conditions of full compliance in the Rawlsian vein) given one kind of project (say, trying to get at general normative principles that govern fair cooperation in a pluralistic society) won't always translate over to *other* idealizations for *other* projects. If we think of all of the projects as instances of "ideal theory", we run the risk of thinking that they must all share a rationale, and thereby ignoring the justificatory structure relevant to our specific project.

Another kind of danger arises *among* inquirers. As we have seen, political inquirers frequently use the term 'ideal theory' to pick out *different* idealizing assumptions. Given this context, there is a clear worry that framing discourse in terms of "ideal theory" (or "non-ideal" theory) may obscure the different idealizing assumptions that different speakers are (or are not) making in a particular context. This lack of clarity in turn may render communication using the terms less fruitful.

We can pull all of these strands together to summarize our case against retaining 'ideal theory' as follows: 'ideal theory' and its cognates are technical terms, but lack the virtues that characteristically warrant the use of such terms. The pattern of usage for these terms may suggest reference failure or polysemy. The best-case interpretation—that the term is context-sensitive—is an unattractive feature for a technical term in this context. And the actual patterns of use raise dangers of unreliable inferences and confusion that we should seek to avoid with our technical terminology. We take these considerations to constitute a strong prima facie case against retention.

5.6 The Case Against Reform

In this section, we argue against the idea that we should seek to *reform* existing usage of 'ideal theory' and its cognates rather than *abandon* use of the terms. Many inquirers discussing "ideal theory" and "non-ideal theory" state how they want their usage of 'ideal theory' and its cognates to be read in the context of their work. These suggestions might well be used as the basis for a potential broader reform of the terminology beyond the context of just their work. We cannot address every possible reforming proposal here. Instead, we first offer general reasons for pessimism about useful reforms, and then explain our pessimism about what we take to be two of the most attractive reform proposals.

We take the functional heterogeneity data canvassed in Sect. 5.4 to cast doubt on the general possibility of an attractive reform. The core reason is as follows. We think it is plausible that *certain* idealizations can sometimes be epistemically useful for each of the three reasons discussed in Sect. 5.4, in some contexts in political inquiry. But we think that it is highly implausible that there is a *specific collection* of idealizations that is useful across a *wide range* of topics in political inquiry, in a *wide range* of contexts. For example, we think that it is implausible that a specific collection of idealizations (each to a particular degree) is *constitutive* of any wide range of important normative categories of interest to political inquiry.[49] We think similar skepticism is warranted about there being a unified pattern of idealizations that are explanatorily or epistemically useful.

If this is so, then any specific way of reforming the content of 'ideal theory' looks like it will likely produce either a gerrymandered mess (if it adequately captures the range of central roles for idealizations), or a theoretically dubious ad hoc cut (if it does not). These issues suggest at first blush that reformed 'ideal/non-ideal' language would not be epistemically useful to inquiry in political inquiry.

The same points hold if we focus on the three ways that idealizations can be unhelpful or pernicious (discussed in Sect. 5.4). Again, it is very plausible that *certain* idealizations can sometimes be theoretically unhelpful or pernicious in each of these three ways. But the diversity of idealizations and how they interact renders it implausible that there is a non-gerrymandered way of reforming 'ideal theory' such that this term can express an "alertive" concept that enables us to reliably avoid these dangers. We could, of course, try to reform 'ideal theory' so that it refers to all and only the theories that involve pernicious idealizations. But, again, this approach would likely be gerrymandered and unmotivated.

This discussion suggests that there isn't a distinctively theoretically useful cut in the vicinity of the usage of 'ideal theory' (or: 'non-ideal theory') for us to regiment to. We think that this result is unsurprising, given the heterogeneous use data: if there really were a single theoretically important cut, we would expect competent inquirers to eventually converge upon it. Instead, what we see in the recent literature is, if anything, an accelerating proliferation of cross-cutting usages.

We can complement this argument with a practical argument about the asymmetry between reform and abandonment. Suppose an individual chooses to abandon the use of a term. She might well want to recruit others to the cause—perhaps the more the better. But even if she does not get others to go along with her, she might

[49] This is a natural place for certain philosophers (e.g., orthodox Rawlsians) to resist our argument, given other commitments they have.

still reap many of the benefits of abandonment in her own conversations: for example, both she and her interlocutors might avoid the kinds of "unreliable inference" worries we glossed above.

Contrast this situation with seeking to reform the use of a term. An inquirer might want to only reform the use (or meaning) of a term as it is used in a given context: for example, its use in their own article. (Perhaps the urge to do this is part of what explains the proliferation of competing characterizations of 'ideal theory' and its cognates.) Even that arguably takes *somewhat* more work to pull off than abandonment, and comes with more risks (e.g., continued vulnerability to the sorts of "unreliable inferences" mentioned in the previous section).

Moreover, many who seek reform want to reform the meaning (or use) of a term more generally, beyond just their own usage. That requires getting a critical mass of people to go along with the reform. To significantly reform the use of 'ideal theory' in this more ambitious way would, at the very least, require the investment of significant resources. In some cases, this sort of investment might be worthwhile. For example, if one thinks that a certain reform to existing patterns of use of race and gender terms would have non-trivial liberatory payoffs, those potential payoffs might warrant considerable investment into bringing about that reform.[50] But in the case of 'ideal theory' and its cognates, to put it bluntly, why would we bother doing this, if (as we have been arguing) the theoretical payoffs look so meager?

We now complement these general points with a discussion of two specific possible reforming proposals. The first is to simply reform usage so that it conforms with the introductory stipulations offered by Rawls. There are three reasonable motives for focusing on this reform. First, the relevant usage is familiar to participants in the discourse: almost everyone in contemporary political inquiry knows about Rawls' stipulations. Second, given Rawls' historical influence, one might hope that this reform would preserve continuity with a large fragment of the extant literature. Finally, on some views about language, one might insist that Rawls' successful introduction attempt *fixed the meaning* of the terms, and so this sort of reform to our usage would (desirably) lead participants to stop using these terms *incorrectly*.

Our core objection to this proposal is that the discussion in Sect. 5.4 suggests reasons to doubt that there is anything especially useful about Rawls's terminology. That is, it is far from clear that using the term 'ideal theory' to group together theories that assume full compliance, and circumstances hospitable to the realization of a certain conception of justice, carves the space of political views at its joints. When we combine this uncertainty with our general point above about the costs of attempting reform, we find it implausible that it is worth investing resources in trying to turn the clock back to have the meaning of 'ideal theory' and its cognates line up precisely with Rawls's original usage. And this is especially true because—without reform—one can make it perfectly clear that one wants to talk about "ideal theory

[50] We take this to be a reasonable reading of a key part of the rationale for Haslanger's work in Haslanger (2000).

in the sense spelled out in *A Theory of Justice*". As this makes clear: abandoning the terminology does not make the Rawlsian *idea* (or any of the other ideas that have traveled under this banner) unavailable to theorists.

A second proposal is to reform to something broader than Rawls's specific proposal, but which—one might think—captures the *genera* of which Rawls' theory is a species. To illustrate, consider what we take to be one of the most promising versions of this idea: to regiment to the idea that an "ideal theory" of x is a theory of something that is *perfectly x*. To see the basis for this idea, consider Rawls' gloss on ideal theory as providing an account of a "perfectly just" basic structure of society.[51] The proposal on the table involves a genera notion, because different inquirers might disagree about (e.g.) what sorts of idealizations are constitutive of perfect justice. On this reform, such theorists would be understood as engaging in substantive disputes within "ideal theory".

We have three worries about this sort of proposal. First, it is not clear that this proposal has the sort of theoretical virtues that would warrant efforts to reform existing usage. Second, it is not at all clear that inquirers who build different idealizations into their notion of "perfect justice" are in fact engaged in substantive disputes about what perfection in justice consists in. For someone might be right that x is just relative to idealization A, and someone else might be right relative to idealization B. What is not clear is that there is anything of substance to the idea that idealization A—and not idealization B—is constitutive of *perfection* in justice. Third, recall an earlier point we made: on many contemporary usages, "non-ideal" theory, is not just any kind of theory that isn't "ideal theory". Rather, theorists posit a variety of competing specific relations between "ideal" and "non-ideal" theory. So, to be complete, this proposed regimentation of 'ideal theory' and its cognates would need to take a stand on which of these ways to regiment use of the term 'non-ideal'. But again, it is not clear that there is any distinctive theoretical payoff for such regimentation, which would warrant the effort needed to successfully reform usage.

5.7 Conclusion

In this paper, we've given an initial case for abandoning using the term 'ideal theory' and its cognates in the context of political inquiry. We can briefly sum up the case in this way. Start with the fact that the existing use of these terms is characterized by an alarming range of cross-cutting ideas. Further, we see a broad range of idealizations that might be useful to political inquirers, and a diverse range of virtues and vices that such idealizations can exhibit across different contexts. On this basis, we have argued against A) *retaining* the existing pattern of use (because it exhibits significant vices, and lacks the characteristic virtues of technical terminology), and also B) seeking to *reform it* (because there is nothing theoretically

[51] Rawls (1971/1999, p. 216).

important to associate with these words that warrants the necessary investment of resources to get inquirers to converge on a regimented pattern of use).

The conclusion of our main argument raises a natural question: if we are to cease using "(non-) ideal theory" talk, how should we be talking about the relevant issues instead?

In our view, there is an attractive way forward after abandonment. When engaging in political inquiry, we should focus our attention on the *particular* idealizations that are being made in a given argument, theory, etc.—or, somewhat more broadly, the particular *kinds* of idealizations. As part of this endeavor, we can focus on which dimensions of the context under consideration are being idealized—e.g., whether the idealizations concern the compliance of agents with institutions and laws, the emotions of agents, the availability of natural resources, etc. We can also make explicit *how* much idealization we are assuming on a given dimension. With this specificity of focus, it will then be natural to speak directly to *why* certain idealizations (and not others) are being made. And it will also (often) be natural to explain why those idealizations are distinctively relevant to our inquiry, why they are not vulnerable to the problem of the second best, and why they are not ideologically pernicious. There would be no illusion that, by using the term 'ideal theory' (or 'non-ideal theory'), we were somehow in a position to assume that these questions had already been answered.

Doing these things, we claim, holds the promise of avoiding all of the main issues we introduced above for contemporary discussion of "ideal" and "non-ideal" theory, or at least doing better on this front than we could do by continuing to use this terminology. Further, given the heterogeneity of the existing literature, and that of the theoretical function of idealizations, we don't see what of value would be lost on our alternative approach besides a certain kind of discontinuity with the existing literature. Especially given how heterogeneous the use of 'ideal theory' and its cognates is in the existing literature, it's not clear if that is a meaningful loss at all.

In closing, it is worth (re-)emphasizing and developing a point we made in the Introduction. Our argument for abandoning "ideal theory" talk is orthogonal to the evaluation of the practice of idealization in political inquiry more generally. To take the case of the central term 'ideal theory': we are arguing for abandoning the use of that term, not against the use of *idealization* in political inquiry, or against doing the kind of inquiry that some characterize as "ideal theory". We think that many of the debates that currently travel under the banner of "ideal vs. non-ideal theory" involve issues of substantial theoretical importance. We also think that theories routinely characterized using 'ideal theory' and its cognates can be worth investigating and developing. We just think that it would be better to talk about these topics, and pursue these projects, with terminology that does not suffer the vices we have identified for 'ideal theory' and its cognates.

Let us illustrate this point. We are sympathetic to many of the dangers of idealization sketched in Sect. 5.4. These dangers are often used as the basis for criticisms of the project of "ideal theory". And an important recent trend in political inquiry involves quick dismissals of all "ideal theory" as *unnecessary* for normative guidance, *irrelevant* to the evaluation of our actual circumstances, objectionably *ideological*, etc.[52] The points we have made in this paper suggest reasons for skepticism about these conclusions, when they are understood to target all political inquiry that makes use of idealizations.

We grant that each of the dangers of idealization has merit when applied to certain specific patterns of idealization. But the functional heterogeneity data suggests that we should be skeptical that these dangers apply to the use of idealization in political inquiry quite generally. In light of this, we take blanket dismissal of "ideal theory" to be another example of a kind of unreliable inference, invited by the illusion of unity produced by organizing inquiry around a theoretically unhelpful technical term.

Obviously, we haven't given anything like a conclusive argument for abandoning 'ideal theory' and its cognates in this short paper. But we hope that our argument is powerful enough to give pause. If it succeeds in prompting readers to think more about why we are giving this terminology a central place in our thought and talk about politics, we'll take that as a win.

Acknowledgements Thanks to Sonu Bedi, Herman Cappelen, Andy Egan, David Enoch, David Estlund, Sean Kim, Zoë Johnson King, Zachary Lang, Russ Muirhead, Timothy Rosenkoetter, Adrian Russian, Zoe Thierfelder, Amie Thomasson, the editors of this volume (Manuel Gustavo Isaac, Steffen Koch, and Kevin Scharp,) and anonymous referees for helpful feedback and discussion. Earlier versions of this paper were presented at Dartmouth College, University of Toronto, and Hong Kong University. Thanks to everyone who participated in those discussions.

References

Anderson, E. (2010). *The imperative of integration*. Princeton University Press.

Berker, S. (2019). The explanatory ambitions of moral principles. *Nous, 53*(4), 904–936.

Burgess, A., & Plunkett, D. (2013a). Conceptual ethics I. *Philosophy Compass, 8*(12), 1091–1101.

Burgess, A., & Plunkett, D. (2013b). Conceptual ethics II. *Philosophy Compass, 8*(12), 1102–1110.

Burgess, A., & Plunkett, D. (2020). On the relation between conceptual engineering and conceptual ethics. *Ratio, 33*, 281–294.

Burgess, A., Cappelen, H., & Plunkett, D. (2020). *Conceptual engineering and conceptual ethics*. Oxford University Press.

Cappelen, H. (2013). Nonsense and illusions of thought. *Philosophical Perspectives, 27*(1), 22–50.

Cappelen, H. (2023). *The concept of democracy: An essay on conceptual amelioration and abandonment*. Oxford University Press.

Cappelen, H., & Dever, J. (2021). On the uselessness of the distinction between ideal and non-ideal theory (at least in the philosophy of language). In J. Khoo & R. K. Sterken (Eds.), *Routledge companion to social and political philosophy of language* (pp. 91–106). Routledge.

Cappelen, H., & Plunkett, D. (2020). A guided tour of conceptual engineering and conceptual ethics. In A. Burgess, H. Cappelen, & D. Plunkett (Eds.), *Conceptual engineering and conceptual ethics* (pp. 1–26). Oxford University Press.

Carnap, R. (1947/1956). *Meaning and necessity: A study in semantics and modal logic*. University of Chicago Press.

[52] See Mills (2005) for an important statement of these kinds of concerns about "ideal theory".

Chalmers, D. (2020). What is conceptual engineering and what should it be? *Inquiry*.

Cohen, G. A. (2003). Facts and principles. *Philosophy and Public Affairs, 31*(3), 211–245.

Cohen, G. A. (2008). *Rescuing justice and equality*. Harvard University Press.

Enoch, D. (2018). *Against utopianism: Noncompliance and multiple agents* (p. 18). Philosophers' Imprint.

Geuss, R. (1981). *The idea of a critical theory: Habermas and the Frankfurt School*. Cambridge University Press.

Haslanger, S. (2000). Gender and race: (what) are they? (what) do we want them to be? *Nous, 34*(1), 31–55.

Hooker, B. (2023). Rule consequentialism. In E. N. Zalta & U. Nodelman (Eds.), *The Stanford encyclopedia of philosophy* (Spring 2023 Edition). https://plato.stanford.edu/archives/spr2023/entries/consequentialism-rule/

Isaac, M. G. (2020). How to conceptually engineer conceptual engineering? *Inquiry*, 1–24.

Jackson, F. (1998). *From metaphysics to ethics: A defence of conceptual analysis*. Clarendon.

Jaggar, A. M., & Tobin, T. W. (2013). Situating moral justification: Rethinking the mission of moral epistemology. *Metaphilosophy, 44*(4), 383–408.

Lipsey, R. G., & Lancaster, K. (1956–1957). The general theory of second best. *The Review of Economic Studies, 24*(11), 11–31.

McKean, B. (2017). Ideal theory after Auschwitz? The practical uses and ideological abuses of political theory as reconciliation. *Journal of Politics, 79*(4), 1177–1190.

McPherson, T., & Plunkett, D. (2017). The nature and explanatory ambitions of metaethics. In T. McPherson & D. Plunkett (Eds.), *The Routledge handbook of metaethics* (pp. 1–25). Routledge.

McPherson, T., & Plunkett, D. (2020). Conceptual ethics and the methodology of normative inquiry. In A. Burgess, H. Cappelen, & D. Plunkett (Eds.), *Conceptual engineering and conceptual ethics* (pp. 274–303). Oxford University Press.

McPherson, T., & Plunkett, D. (2021). Conceptual ethics, metaepistemology, and normative epistemology. *Inquiry 68*(7), 1501–1533.

McPherson, T., & Plunkett, D. (2024a). Normative standards and the epistemology of conceptual ethics. *Inquiry, 67*(3), 954–984.

McPherson, T., & Plunkett, D. (2024b). Topic continuity in conceptual engineering and beyond. *Inquiry, 67*(9), 2847–2873.

Mills, C. W. (2005). "Ideal theory" as ideology. *Hypatia, 20*(3), 165–184.

O'Neill, O. (1996). *Towards justice and virtue: A constructive account of practical reasoning*. Cambridge University Press.

Olson, J. (2014). *Moral error theory: History, critique, defence*. Oxford University Press.

Pinder, M. (2021). Conceptual engineering, metasemantic externalism and speaker-meaning. *Mind, 130*(517), 141–163.

Plunkett, D. (2016). Conceptual history, conceptual ethics, and the aims of inquiry: A framework for thinking about the relevance of the history/genealogy of concepts to normative inquiry. *Ergo, 3*(2), 27–64.

Plunkett, D. (2024). The varieties of idealization and the politics of economic growth: A case study on modality and the methodology of political philosophy. *Inquiry, 67*(6), 1908–1946.

Rawls, J. (1971/1999). *A theory of justice* (revised edition). Harvard University Press.

Rose, J. L. (2020). On the value of economic growth. *Politics, Philosophy and Economics, 19*(2), 128–153.

Scharp, K. (2013). *Replacing truth*. Oxford University Press.

Sen, A. (2006). What do we want from a theory of justice? *Journal of Philosophy, 103*(5), 215–238.

Shelby, T. (2016). *Dark ghettos: Injustice, dissent, and reform*. Harvard University Press.

Simmons, A. J. (2010). Ideal and nonideal theory. *Philosophy and Public Affairs, 38*(1), 5–36.

Smith, M. (1994). *The moral problem*. Blackwell.

Stemplowska, Z. (2008). What's ideal about ideal theory? *Social Theory and Practice, 34*(3), 319–340.

Stemplowska, Z. (2017). Non-ideal theory. In K. B. Kasper Lippert-Rasmussen & D. Coady (Eds.), *A companion to applied philosophy* (pp. 284–296). Wiley.

Stemplowska, Z., & Swift, A. (2012). Ideal and non-ideal theory. In D. Estlund (Ed.), *The Oxford handbook of political philosophy* (pp. 373–388). Oxford University Press.

Strawson, P. F. (1963). Carnap's views on conceptual systems versus natural languages in analytic philosophy. In P. A. Schilpp (Ed.), *The philosophy of Rudolf Carnap* (pp. 503–518). Open Court.

Tessman, L. (2009). *Feminist ethics and social and political philosophy: Theorizing the non-ideal.* Springer.

Thomson, C. (2020). Ideal and nonideal theory in political philosophy. In *Oxford research encyclopedia of politics*. Oxford University Press.

Valentini, L. (2012). Ideal vs. non-ideal theory: A conceptual map. *Philosophy Compass, 7*(9), 654–664.

Wiens, D. (2020). The general theory of second best is more general than you think. *Philosophers' Imprint, 20*(5), 1–26.

Chapter 6
Representing or Shaping Reality? What *Class* Can Teach About *Woman*

Teresa Marques

Abstract Haslanger (Noûs 34(1):31–55, 2000) has argued that we should ameliorate concepts of race or gender to better capture existing structural inequalities. Her analysis was criticized by Simion (Inquiry 61(8):914–928, 2018a), who argued that a concept should be ameliorated only if doing so preserves epistemic accuracy. But, as I argue, this criticism misses Haslanger's target. In response, Podosky (Inquiry 1–15, 2018) and McKenna (Logos Episteme 9(3):335–342, 2018) have argued that conceptual revisions need not preserve "epistemic accuracy" since concepts can "shape reality", not just represent it. Here I argue that social kind concepts need not "shape" the corresponding social kinds: neither causally nor constitutively. This raises doubts about the function of ameliorative projects for social kind concepts as means to change social reality.

Keywords Conceptual amelioration · Epistemic accuracy · Social class · Woman · Social kinds

6.1 Introduction

Sally Haslanger has argued that we should analyse social kind concepts for discriminated groups in a way that lays down the existing conditions of subordination of the relevant social kinds. In particular, she has suggested that concepts for race and gender could be ameliorated to reveal specific forms of structural inequality. These are primarily *theoretical* aims and meeting them is a required precondition for understanding and addressing inequality. Haslanger's project thus follows what

T. Marques (✉)
LOGOS Research Group, BIAP – Barcelona Institute of Analytic Philosophy, Universitat de Barcelona, Barcelona, Spain
e-mail: teresamatosferreira@ub.edu

M. G. Isaac et al. (eds.), *New Perspectives on Conceptual Engineering - Volume 2*, Synthese Library 514, https://doi.org/10.1007/978-3-031-98524-9_6

historically came to be known as *materialist feminism*. Since Simone de Beauvoir, materialist feminists tried to draw from the Marxian analysis of work, subordination, and class to describe the enduring inequalities affecting women.

Over the last 20 years, Haslanger's analyses of race and of gender gave rise to numerous debates. It is not my goal here to engage all of them.[1] I'm also not interested in doing Haslangerian hermeneutics. Rather, I intend to focus on some claims about the epistemology of conceptual amelioration and some of the related metaphysics of social kinds. Simion (2018a) has argued that a concept should be ameliorated only to the extent that doing so preserves epistemic accuracy. However, she has also argued that Haslanger's concept *woman* is not epistemically accurate, because it does not represent all women (Simion, 2018b). Her arguments for an epistemic constraint on amelioration have been contested. Podosky (2018) claims that conceptual revision should be allowed whenever it has the "capacity to causally influence the world". McKenna (2018) has similarly argued that concepts can "shape reality" and hence that there is no epistemic loss in ameliorating a concept that is not accurate *right now.* A common assumption in the literature seems to be that ameliorative analyses necessarily *revise* concepts to bring about social change. Perhaps this can be interpreted in different ways. The sense that will concern me here is one where conceptual revisions of social kind concepts serve to change or create new social kinds for which revised concepts would stand.

Here, I argue that conceptual ameliorations do not necessarily *revise* existing concepts, and, more importantly, that ameliorative analyses are better understood as *introducing* new concepts to better understand *existing* social kinds. Arguably, this is what Haslanger (2000) ameliorative project for gender and race concepts tried to do. But the concept of gender as a social class is not a modification of the pre-theoretical folk concept of women as *human females,* arguably a concept expressed across all languages. It is a different, albeit related, concept.

This has broader implications. The fact that new concepts are needed to capture preexisting social kinds shows that having an ameliorated concept is *not necessary* for the existence of the corresponding social kind. On the other hand, some social kind concepts may not pick actual, desirable, or even possible kinds. Therefore, social kind concepts may not "shape" social kinds: neither causally nor constitutively.

A hypothetical criticism of Haslanger's analyses of race and gender concepts would have to show that it failed for other semantic, epistemic, or practical reasons related to its aims, say, that it did not accurately lay down the conditions of the social and historical subordination of human females or of different racialized peoples. Alternatively, it could question if theorizing about subordination based on sex or race requires introducing new senses for old words. Alas, the discussions I engage with in this paper do not make such critiques.

[1] This profuse literature includes, for example, Andler (2017), Bach (2012, 2016), Barnes (2020), Bogardus (2020), Jenkins (2016), Laskowski (2020), Mikkola (2009), Mills (2014), Saul (2006).

This paper is organized as follows. In Sect. 6.2, I summarize Simion's criticism of Haslangerian ameliorative projects, and Podosky's and McKenna's responses to Simion. This will serve to raise two important related questions in conceptual engineering: whether ameliorative projects *revise* existing concepts or *introduce* new concepts, and whether concepts *shape* reality, either causally or constitutively. In Sect. 6.3, I argue that, first, social kind concepts are not sufficient for "shaping" social kinds, either causally or constitutively, particularly when they are "ideological". Second, Simion's criticism does not really apply to Haslanger's account. Finally, in Sect. 6.4, I argue that ameliorative analyses need not revise (modify) existing concepts, and that insofar as they devise new social kind concepts, as Haslanger did for gender and race concepts (in the tradition of the Marxian analysis of social classes), social kind concepts are also not necessary to "shape" the intended corresponding social kinds, neither causally nor constitutively.

6.2 The Requirement of Epistemic Accuracy in Conceptual Engineering

6.2.1 Representing Reality

When should a concept be 'engineered', i.e., revised or ameliorated? Mona Simion (2018a) is optimistic about the prospects of conceptual ameliorations but considers that they should be subject to epistemic constraints. A concept *should* be ameliorated, Simion claims, only insofar as doing so does not entail epistemic loss. This may pose a problem for some proposals of ameliorative projects that have primarily political or moral aims.

One such salient proposal can be found in the work of Sally Haslanger. In her (2000) paper on gender and race, Haslanger stated four concerns that an account of gender and race as social classes should consider and set out to address two of those concerns, namely:

(i) The need to identify and explain persistent inequalities between females and males, and between people of different "colours".
(ii) The need for a framework that will be sensitive to both the similarities and differences among males and females, and the similarities and differences among individuals in groups demarcated by "colour".

Haslanger further introduces three principles of analysis that a materialist theory of gender and race as social classes should adopt. First, gender categories should be defined in terms of how one is socially positioned. Second, these categories should be defined hierarchically within a broader complex of subordinating relations. Finally, sexual differences are revealed to be the physical markers that distinguish the members of the two groups and are used in justifying the differential treatment of the members of each group.

Haslanger thus lays down a way of understanding gender as a social class that is compatible with the tradition of materialist feminism. Her proposed analyses of the social classes of gender are now well known:

> *S is a woman* iff$_{df}$ S is systematically subordinated along some dimension (economic, political, legal, social, etc.), and S is "marked" as a target for this treatment by observed or imagined bodily features presumed to be evidence of a female's biological role in reproduction.
>
> *S is a man* iff$_{df}$ S is systematically privileged along some dimension (economic, political, legal, social, etc.), and S is "marked" as a target for this treatment by observed or imagined bodily features presumed to be evidence of a male's biological role in reproduction. (Haslanger, 2000, p. 39)

Now, Simion takes issue with revisionary purposes that (may seem to) neglect concepts' primary function (Simion, 2018b, p. 95):

> [I]n virtue of their main, representational epistemic function, concepts will be properly functioning when responsive to epistemic reasons pertaining to properly representing the world, and malfunctioning when merely responsive to other types – practical, moral, political.[2]

And so, amelioration motivated by the pursuit of practical or political aims may suffer from a 'wrong-kind-of-reasons' problem if it neglects the main representational function of concepts.[3]

The wrong-kind-of-reasons problem can be summarized in the following argument: The primary or etiological function of concepts is a representational epistemic function. Revising or changing a concept for purposes that defeat their primary representational function entails epistemic loss, that is, loss of epistemic accuracy. Therefore, conceptual engineering for non-primary function reasons that result in epistemic loss amount to engineering for the wrong kind of reasons (Simion, 2018a, pp. 6–7). Thus, a concept should be engineered only if does not result in epistemic loss. In other words, conceptual engineering is constrained by an epistemic principle:

> *The Epistemic Limiting Procedure (ELP)*: A representational device should be ameliorated iff (1) There is all-things-considered reason to do so and (2) The amelioration does not translate into epistemic loss. (Simion, 2018a, p. 10)

Simion seems to implicate that our term 'woman' is "semantically perfectly fine" and that the concept it expresses does "exceptionally well" epistemically, i.e., it "carves nature at its joints" (Simion, 2018a, p. 7). In Simion (2018b), she argues that Haslanger's proposed amelioration of *woman* does entail epistemic loss, because the proposed analysis leaves some women out. However, if the main function of *woman* is to pick out women, a concept that fails to pick out what it should

[2] In Simion and Kelp (2019) they talk rather of etiological functions.

[3] Isaac (2024) has questioned Simion's contention that the amelioration of ideological concepts should be constrained by epistemic accuracy. See also the reply in Podosky (2022). For more general overview on the literature on conceptual variability and conceptual change, see for instance Marques and Wikforss (2020).

represent has failed in its main epistemic function and hence malfunctions. Thus, it's not a good concept:

> If Haslanger's 'woman' fails to be a good concept *qua* concept, plausibly, it will not be a better concept than its predecessor. If so, Haslanger's project will fail to qualify as an ameliorative project: it will not have engineered better ways for us to think about the world. (Simion, 2018b, p. 97)

That is harsh, and as I will soon argue, misses its target. Let us recapitulate the argument:

1. Assuming *ELP*, we should engineer a concept just in case there are reasons (such as political reasons) to do so, and there is no epistemic loss in it.
2. If we engineered the concept of woman in the way suggested by Haslanger, there would be epistemic loss, since there are women who don't fall under Haslanger's analysis.
3. Therefore, we should not engineer the concept of woman in the way suggested by Haslanger.

I will show that Simion is mistaken when she claims that Haslanger's analysis entails epistemic loss, and thus reject (2).

Haslanger (2000) recognized that what she was then advancing was

> …an analytical approach to defining race and gender. However, its analytical objectives are linked to the descriptive project of determining whether our gender and race vocabularies in fact track social kinds that are typically obscured by the manifest content of our everyday race and gender concepts. Although the analyses I offer will point to existing social kinds and this is no accident, I am not prepared to defend the claim that these social kinds are what our race and gender talk is "really" about. My priority in this inquiry is not to capture what we do mean, but how we might usefully revise what we mean for certain theoretical and political purposes. (Haslanger, 2000, p. 34)

She makes clear in this passage that her objective is to give a descriptive project, that is, an epistemically sound project, and not to "capture what we do mean".

6.2.2 Shaping Reality

There are several rejoinders to Simion's argument. Paul-Mikhail Podosky (2018) proposed an improvement on *ELP* and drew attention to the possibility that ameliorated concepts can *come to be* accurate in the future. Epistemic loss, as it stands, is not an adequate criterion since it "does not give us all-things-considered epistemic reason to refrain from amelioration in cases where the revised concept has the capacity to causally influence the world, and can make itself representationally accurate." (Podosky, 2018, p. 11). Moreover, he suggests, conceptual engineering is often "about" epistemic loss:

> This is because ideology serves to create social reality; it makes certain facts true. And ideology consists of concepts. Hence, ideological concepts are accurate because they *constitute social reality*. Given that ideology is oppressive, then it appears that primacy should

> not be given to knowledge preservation. Despite the accuracy of ideological concepts, we should not preserve them if they perpetuate oppression, no matter the epistemic cost. (Podosky, 2018, p. 2. My emphasis)

Later he adds: "ideology partly constitutes social reality. Thus, ideological concepts accurately represent social reality." (Podosky, 2018, p. 8)

Podosky's defence of provisional epistemic loss appears at first sight to be in line with Haslanger's claim that "it is part of the project of feminism to bring about a day when there are no more women" (Haslanger, 2000, p. 46). But, as I'll argue in the following sections, this interpretation is misguided. Concepts themselves are neither necessary nor sufficient to either cause or constitute the relevant features of social reality, i.e., the corresponding social kinds. It is unclear whether the pre-theoretical folk concept is ideological, and neither the pre-theoretical folk concept nor Haslanger's ameliorated concept *woman* cause or constitute social reality (in any plausible sense: bring about, create, cause, or make it the case that women are systematically discriminated), as I'll argue in the following sections.

McKenna (2018) makes a similar point to Podosky's and claims that Simion's objection fails because social kind concepts "have the primary function of shaping the world" (McKenna, 2018, p. 341). This would mean that gender (and race) concepts, as social kind concepts, serve to *shape* the world, not (merely) t*o represent it.* McKenna very briefly explains what this means: "Social kind concepts like 'husband' and 'wife'... refer to social roles that are in part created and maintained by our practices involving them." (McKenna, 2018, p. 340) What it is to be a husband, or a wife, depends on social conventions and norms, and on the social practices regulated by these norms. For two people to become husband and wife is not for them to be *caused* to be married. It is, rather, to come to occupy a social position in the context of a given conventional practice. Does that mean that ideological concepts can *create* social reality?

Both of Podosky and McKenna's replies to Simion—emphasizing the impact of ideology on reality, should be interpreted as continuous with a tradition of Marxian thought and of the impact of ideological beliefs, through human conduct, in the world. To address their claims, I will bring in the background of these Marxian influences on materialist feminism.

6.3 Concepts Are Not Sufficient for "Shaping" Reality

6.3.1 The Rejection of Idealism in Marxian Theory

The question that gives the title to this paper—*representing or shaping reality?*—is the contemporary reflection of a historical debate between materialists and idealists. All parties can agree that thought and language represent reality. However, the view that ideas, concepts, or representations, shape the world may be seen as the contemporary version of idealism, which Marx opposed. As Jonathan Wolff (1992) explains, Marx rejected two philosophical traditions: ahistorical materialism (represented in authors like Hobbes or Feuerbach), which sees the human mind as inertly representing what is outside of it, and Hegelian historical idealism, the thought that the mind can construe reality and that a change of minds would bring about changes in the world.

> The root idea – one Marx finds neglected in all previous philosophies – is that human beings have needs, and need, not contemplation, is their primary relation to the world. (Wolff, 1992, p. 38)

Marx's alternative to these two philosophical traditions is to understand our interaction with the world as a practical activity. If he was right—that human needs and their satisfaction are how people primarily relate to the world—then the mere revision of ideas (concepts, representations) in and of itself will not change *human needs*. Indeed, changing word meanings, or changing people's ideas, may come to *obscure* reality, while not changing the underlying human needs. For all we know, human nature may impose constraints on realistic possibilities for social change. Ironically, some failed historical attempts to go against "human needs" were motivated by Marxist ideology, for instance, attempts to abolish private property[4] or the family.[5]

In correspondence with Marx, Engels introduced the phrase *false consciousness* while characterizing ideology:

> Ideology is a process accomplished by the so-called thinker consciously, it is true, but with a false consciousness. The real motive forces impelling him remain unknown to him; otherwise it simply would not be an ideological process. Hence he imagines false or seeming motive forces. (Marx and Engels, 1975, 1975 Selected Correspondence, p. 459)

In her book on false consciousness, Meyerson (1991) addressed some interpretations of Marxism to the effect that, first, ideologies play a role and have important *causal effects* on our lives, and second, that because of that, they *partially constitute* reality. Against the first point, Meyerson emphasizes that *the fact that ideological beliefs reflect an observer's experience doesn't show them to be true*. Against the second, Meyerson remarks that *the fact that a belief has causal effects or plays a role in the explanation of people's conduct is compatible with that belief being false*. The problem with such interpretations of Marxian thought, she says, is "their implicit idealism, their assumption that the experienced world is the real world or that mental states like beliefs can somehow constitute reality" (Meyerson, 1991, p. 5).

Now, Meyerson provides her claim in the context of an exegetical interpretation of the writings of Marx and Engels. But her points are easily illustrated. In her book *The Witches: Salem, 1692 A History*, Stacy Schiff (2015) writes:

---

[4] See the discussion about the long-term financial consequences for Cambodia of the Khmer Rouge's abolition of money, markets and Cambodia's Central Bank (Prasso, 2001).

[5] See Glass and Stolee (1987) about attempts to abolish family relations after the Bolshevik revolution in the USSR.

> In 1692 the Massachusetts Bay Colony executed fourteen women, five men, and two dogs for witchcraft… The town's senior minister discovered he was related to no fewer than twenty witches. Ghosts escaped their graves to flit in and out of the courtroom, unnerving more than did the witches themselves. (Schiff, 2015, pp. 3–4)

The Salem witch trials have all the markings of social phenomena caused by ideological belief. They also nicely illustrate the two points that Meyerson makes. First, *the fact that ideological beliefs reflect an observer's experience doesn't show them to be true*—people may have been unnerved that "ghosts escaped their graves to flit in and out of the courtroom", but ghosts and witches do not exist. Indeed, what we refer to with the phrase "witch trial" is *not* an apt description of the prosecution of its victims, since there are no witches. Second, the fact that a belief has causal effects or plays a role in the explanation of people's conduct is compatible with that belief being false. False beliefs can explain conduct and have important causal effects on those falsely accused. Although "in all of New England it would have been difficult to find more than a few souls to whom the supernatural was not eminently real" (Shiff, 2015, p. 8), there is a difference between what is "eminently real" to someone, and what is real.

Ideological prosecutions like witch trials have been recurrent in history. In the Soviet Union, those prosecuted were 'enemies of the people'.[6] But the ideology behind their trials did not make them actual traitors of their motherland. In Marques (2017), I argued that not all possible constitutively socially constructed kinds are actual social kinds. If an existing social group is constitutively constructed as a social kind *K*, the fact that it actually exists as a *K* requires other social causes. The mere existence of a concept for the kind does not entail the existence of the kind.

The background and cases briefly introduced in this section put into question claims to the effect that ideological belief, or ideological concepts, *constitute* the social kinds for which ideological concepts would stand. "Witch trials" names prosecutorial trials of innocent people (and non-human animals, as it turns out), but does not describe trials of witches. The concept *witch trial*, to use Simion's description, malfunctions: it is not responsive to epistemic reasons pertaining to properly representing the world; it is merely responsive to ideological reasons.

Likewise, people accused of being enemies of the people are not thereby *constituted* as members of the group of traitors to the motherland. Obviously, such *autos-de-fé* had ominous consequences.[7] But the *creation* or *constitution* of the presumed

[6] "The disappearance of a father and a husband placed enormous strain on families. Wives and children did not know what to think. Many wives renounced arrested husbands, not necessarily because they thought they might be 'enemies of the people,' but because it made survival easier and gave protection to their families (many husbands for this reason advised wives to renounce them)." (Figes, 2007, pp. 305–307)

[7] For a discussion of cases that pervert, rather than ameliorate, existing concepts, see for instance Marques (2020), Shields (2021), Queloz and Bieber (2021), or Kitsik (2023). It may be suggested that such cases suggest successful conceptual revisions. If that were the case, labeling someone as an enemy of the people in the USSR would have been also symptomatic of a successful revision. This is unwarranted. A project can fail to be ameliorative, while also failing in its aims. As an analogy, consider a policy to reduce immigration, say, the UK leaving the European Union. It may not

corresponding social kinds are not among their effects. This shows that social kind concepts do not straightforwardly *cause* or *constitute* social kinds. They do not shape social reality, and do not suffice for their own accuracy.

6.3.2 *Materialist Feminism and Its Aims*

Haslanger's (2020) analysis followed a long tradition of materialist feminism, itself inspired by Marxian thought. In the introduction of the first volume of *The Second Sex*, Simone de Beauvoir (1949/2010) remarks:

> There have not always been proletarians: there have always been women; they are women by their physiological structure; as far back as history can be traced, they have always been subordinate to men; their dependence is not the consequence of an event of a becoming, it did not *happen*...it falls outside the nature of historical fact. (De Beauvoir, 1949/2009, p. 8)

In a relatively recent talk, Ti-Grace Atkinson (2014) acknowledged the impact of the De Beauvoir's thought on her own:

> The central question for feminists has always been: how does the oppression of women work? Where does it come from and how is it maintained? *We can't dismantle any structures that we do not understand.* Effecting change depends first on our analysis of the problem... I broadened Beauvoir's discussion in order to respond to the question of why, if women were not naturally inferior, our oppression had continued fundamentally unchanged. (Atkinson, 2014, p. 1. My emphasis.)

Haslanger's analysis follows this tradition. She acknowledges that our words and thoughts can *obscure* reality. This does not entail that they constitute reality. Her *ameliorative* analysis of gender and race concepts is meant to *disclose* the social nature of the systemic and structural discrimination of specific classes of people.

> My strategy is to offer a focal analysis that defines gender, in the primary sense, as a social class... As I see it, the core phenomenon to be addressed is the pattern of social relations that constitute the social classes of men as dominant and women as subordinate... As will become clearer below, I see my emphasis as falling within, though not following uncritically, the tradition of materialist feminism. (Haslanger, 2000, p. 37)

Let us then concede to Simion that the proper function of the concept *woman* is to pick out women reliably. Now, there may be different *conceptions* of what women are.[8] At the same time, there may be reasons to think that there exists an externally individuated concept *woman* that picks out women reliably.[9] There are human

---

have been ameliorative, because it was pursued for the wrong reasons (on assumption). But it was also unsuccessful because it failed in its aims, in that it did not reduce immigration.

[8] On the difference between concepts and conceptions, see for instance Sawyer (2020, p. 7).

[9] One could argue that the concept woman is externally individuated by the natural biological kind, in a way similar to how the concept water is externally individuated as the substance with the chemical composition H_2O, after Putnam (1975). (See Byrne (2020)).

females and there are human males, and it is not far-fetched to argue that the folk concept *woman* stands for human females. Here's one way this argument may go.

Barbara Malt (2020) has explained that there used to be a common assumption among psychologists and linguists in the 1950s about how words and concepts could be neatly mapped to external kinds of entities:

> If people actually did perceive the world as falling into natural groupings that demanded distinctive names, we might expect that, to the extent that people around the world share experiences, they would create parallel mappings of those experiences into language...

Now, this assumption has been put into question. Malt's own work focuses on categories for artefacts. She proceeds to show why the assumption is flawed:

> ... the situation is more complicated than that. Things in the world don't always fall into clear groupings, and multiple factors influence the particular distinctions that are labelled by speakers of a given language. Looking across languages, the result is that languages differ from one another in the distinctions they make, choosing different ways of mapping experiences into words. Languages may differ not only in how finely they segment a given domain, but also in what the prototypical referents of the words are, where the boundaries of use lie, and even which dimensions of the domain they encode. (Malt, 2020, p. 22)

The assumption that all things in the world fall into clear groups has thus been rejected. But what about the possibility that there are *some* universal mappings, some concepts that get expressed in all languages? If there were some meanings that are universal across languages, we would have good evidence of concepts for natural groupings that demand distinctive names (Malt, 2020, p. 22).

The Swadesh list is a classic compilation of basic concepts for the purposes of historical-comparative linguistics.[10] The concepts are supposed to be those that have expressions in all languages and to include *woman*, *man* (adult male), *man* (human being or person), *child*, *mother*, among other nouns. It also includes verbs for biological functions like *to eat*, *to breathe*; *to laugh*, *to see*, *to hear*, and even verbs like *to know* or *to think*. There are, in fact, various lists of this kind. Goddard (2002) gave a critical overview of existing studies in typological linguistics that sought to establish any absolute and precise lexico-semantic universal word-meanings. He refers to a 100-word Swadesh list and contrasts it with Wierzbicka's list of universal semantic primes.[11] All languages seem to have words for *something* or *someone* (Goddard, 2002, p. 10). All languages have some word for *people* (Goddard, 2002, p. 12). More directly for our concerns, 'man' and 'woman' have a claim to have universal meanings (Goddard, 2002, p. 14).

Interestingly, the etymology of 'wife', 'woman', and 'man', from Old English, points towards a distinction based merely on sex. 'Woman' (and 'wife') come from

---

[10] There have been some questions about Swadesh lists. Ringe (1992) assessed challenges to the common opinion among mainstream historical linguists that probably human languages are all genetically related, but that the remoter relationships cannot be demonstrated by reliable linguistic methods because the languages in question have diverged too much. The methods in question focus on similarities of distinctive sounds, and matchings of sounds.

[11] Semantic primes or primitives are a set of semantic concepts that are innately understood but cannot be expressed in simpler terms. Semantic primes represent universally meaningful concept. Wierzbicka (1996) provides evidence that all languages use the same set of semantic primes.

'wif'. In Old English and other Germanic languages, 'wif' meant female, and 'man' was the neutral for person, while 'were' means male (as in 'werewolf').[12]

So, we have conceded to Simion that the folk concept of *woman* is "semantically perfectly fine" and does "exceptionally well" epistemically. Even so, Simion's criticism of Haslanger's analysis still misses the target. Of course, Simion is right that perhaps not all women are subordinated. Feminist historians like Gerda Lerner (1986) have argued that the subordination of women started in a historical period (she focuses on Mesopotamia). If she is right, pre-agriculture hunter-gatherer societies would have been more egalitarian (although we may suspect that this isn't quite right).[13]

Why would Simion's argument fail its target, then? It is useful to draw an analogy with the concept *social class*. The *Encyclopaedia Britannica* defines 'social class' as "a group of people within a society who possess the same socioeconomic status." Nearly all societies recognize social classes in this sense. E.g., the great medievalist Norman Cantor (1993) describes how, in the eleventh century, an English monk arriving in Flanders would be surprised to find a new social class:

> The English monk knew of only three social classes—those who fought, those who prayed and those who laboured—but these bourgeois made their living from the manufacture and sale of woollen cloth (Cantor, 1993, pp. 237–238).

Now, different theories can give different explanations of the origin and significance of groups with the same socioeconomic status. Theologians can argue that social classes are the reflection of the spiritual world. Indeed, for centuries the appeal to the "divine right of kinds" was used to justify absolute monarchy in Europe. In India, the caste system stratified society for millennia, and the castes themselves are supposed to have a divine origin (with Brahma, the god of creation. See article 15, *The Constitution of India*).

The meaning of "social class" in Marxian analysis is not merely one of groups with the same socioeconomic status and is not based on theological considerations. Rather, for Marx, class is a function of position in the social network of forces of production and the corresponding *economic relations of power and control*. Class position, in turn, determines ideology and false consciousness, and is conditioned by people's needs and interests. Different social classes are (necessarily?) antagonistic, and their relationship is characterized by class struggle for the control of the means of production. This indicates that the Marxian notion relies on other technical notions deployed in the theory, for instance, that of *forces of production* (a combination of the material means of labour—infrastructure, land, tools, human labour power), *false consciousness*, or *class struggle*.

---

[12] "In the past, it was pronounced *wif*, with the vowel as in Modern Engl. *wee*. It meant "woman," not "female spouse," as it still does in *housewife*, *midwife*, *old wives' tale*, German *Weib*, and Dutch *wijf*. Very early, *man(n)* "person" was added to it, and by a series of phonetic changes *wifman* became *woman*." (Liberman, 2011). Thus, etymologically, the English 'woman' means female human. The plural eventually dropped the 'f', resulting in something like 'wimmin'—closer to the modern plural of 'woman'.

[13] The idea that hunter-gatherers are more egalitarian with respect to gender seems to be based on some "sloppy reasoning". For arguments and evidence as to why that is so, see Kasrnovsky (2021). See also data collected from hunter-gatherer societies in Hayden et al. (1986).

Now, suppose that we want to question the theoretical Marxian analysis of class. The theoretical concept *social class* purports to play an essential function in the diagnosis of the origin and perpetuation of economic inequality, but it is defined with concepts that are not pre-theoretically available. We cannot argue against the Marxian analysis because it differs from common folk pre-theoretical conceptions of social class, or from a dictionary definition. To question the theoretical concept, we must address the terms of its analysis and look for empirical evidence. For instance, we should ascertain whether class struggle for the means of production is an intrinsic or necessary part of social reality.

The contrast between a folk and a theoretical concept is also evident when we shift our attention to other alleged social kind concepts. Haslanger argued that we should adopt social kind concepts for gender and race. In recent work, Machery and Faucher (2020) reported experimental research that supports the claim that people's pre-theoretical race concepts are of biological groups and are constrained by folk biology. Some constraints on shared concepts are external (Machery & Faucher, 2020, p. 169); others are internal and depend on our cognitive constraints. Crucially, biological categories have a "rich inductive potential" (Machery & Faucher, 2020, p. 172). Their results are important to understand how most people actually think about race, but they are mute on the reasons for Haslanger's critique. What she purports to offer are concepts that respond to the theoretical "need to identify and explain persistent inequalities between… people of different 'colors'" (Haslanger, 2000, p. 36). The fact that the folk concept of race is of a biological kind gives no reason, for or against, adopting a social kind concept for the purposes of theorizing about the persistence of racism.

For parity of reason, we can't reject the theoretical analysis of gender as a social kind concept because it differs from the folk concept semantically expressed by the word 'woman' in its pre-theoretical use. Byrne (2020), for instance, has argued that 'woman' means adult human female, thereby defending the folk concept. He gives a series of arguments, including some based on similar reasons considered here: that the folk concept corresponds to the dictionary definition, and that English, like all other languages, would be expected to have a word with such a meaning. The folk concept may even display "the rich inductive purposes" of folk biological concepts. Byrne further considers one argument based on such inductive purposes (Byrne, 2020, p. 3788). But, like Machery and Faucher, he also does not address the issue at hand: that in Haslanger's account concepts for social kinds are introduced for the purposes of social theory, in particular to theorize about the persistence of sexist and racist discrimination. Given this theoretical function, they need not correspond to previously existing operative folk concepts. After all, the Greek for 'individual' is 'άτομο', but contemporary Greek physicists deploy the same concept as physicists when they use 'άτομο' in physics.

Knowing if Haslanger's ameliorative analysis of *woman* tracks any group of people in the world requires empirical research on structural forms of inequality that affect female humans. This evidence is plentiful. In an article for the UN Development Programme, Rivera et al. (2020) report that the COVID-19 pandemic exposed structural inequalities from health to the economy, and from security to social protection.[14] Hard-fought gains for women's rights are also under threat.[15] A report prepared for the United Nations by the human rights group *Walk Free* collected data about disconcerting global trends. Women and girls account for 71% of all victims of modern slavery, 99% of all victims of forced sexual exploitation, 84% of all victims of forced marriage, and 58% of all victims of forced labour. UN Women and the UN Development fund reveal that there have been 23 million sex selective abortions of females, and more than 140 million females are considered "missing" today because of a preference for sons over daughters and of extreme neglect of young girls leading to their death.

A good part of the global economy is based on the exploitation and control of female bodies in ways that remind us of the Marxian concept of alienation (Wolff, 1991, p. 39). The surrogacy industry, the use of (mostly poor) women's bodies, was valued globally at $2.3 billion per year, and is looking to grow into a $27 billion industry by 2025.[16] Profits from human trafficking average $150 billion dollars per year,[17] prostitution $186 billion per year,[18] and pornography $90 billion per year (with $13 billion in the US alone).[19]

In the two previous paragraphs, I used the words "woman", "women" and "girls" with their pre-theoretical meaning. Yet, the paragraphs describe the structural and systemic subordination of women and girls that makes Haslanger's description of gender as a social class plausible: a class that is economically, politically, legally, and socially subjected to systemic inequality.[20]

For clarity, then, we should distinguish between $woman_{HF}$ and $woman_{SC}$, that is, between the folk concept for human females and the theoretical concept for the social class they occupy. The two concepts are obviously related. The social kind concept $woman_{SC}$ picks up "a function of position in the network of relations of power and control", it is the structure that organizes the relative position of $women_{HF}$ and men_{HM} in the social world. This is intended to be *epistemically accurate*.

---

[14] A quick access to the data is available here: https://interactive.unwomen.org/multimedia/explainer/covid19/en/index.html

[15] Detailed information about how Covid 19 impacted women globally and across all dimensions is available at the UN Women site: https://www.unwomen.org/en/digital-library/multimedia/2020/5/explainer-covid19-impacts-women-girls

[16] See https://www.globenewswire.com/news-release/2019/12/16/1960684/0/en/Surrogacy-Market-revenue-to-cross-27-5-billion-by-2025-Global-Market-Insights-Inc.html

[17] See United Nations *Global Report on Trafficking in Persons 2018*: https://www.unodc.org/documents/data-and-analysis/glotip/2018/GLOTiP_2018_BOOK_web_small.pdf and https://www.humanrightsfirst.org/resource/human-trafficking-numbers

[18] http://jamirodman.com/2015/09/30/186-billion-a-year-prostitution-revenue-worldwide-by-country/ and https://apps.urban.org/features/theHustle/index.html and https://www.nytimes.com/2014/03/12/us/in-depth-report-details-economics-of-sex-trade.html

[19] https://fightthenewdrug.org/how-free-porn-industry-growing-in-fast-paced-digitized/

[20] The most recent defence of gender as a social class vs. sex as a biological category, that to my mind is continuous with Haslanger's (2020) analysis of gender and with the tradition of materialist feminism, can be found in the recent work of Lawford-Smith (2022), where she defends the sex/gender distinction, and argues that gender is imposed on people on the basis of sex.

And available data suggests that the concept is, indeed, epistemically accurate. The fact that there are or may be women$_{HF}$ who are not women$_{SC}$ is not a problem for Haslanger's ameliorative project. And hence Simion also appears to miss the point of Haslanger's account.

Could there be other concerns about Haslanger's analysis? Perhaps. For instance, some people could raise objections on the grounds that it does not describe the real basis of structural inequality. As an illustration of how such an argument could be pursued, someone could start by looking at the statistical data biases against women described in Criado Pérez's book *Invisible Women* (2019). For example, only 1 in 8 women who have a heart attack report classic male symptoms such as chest pain, which results in undiagnosed disease in women (Criado Pérez, 2019, p. 197). This is serious, since cardiovascular disease is the leading cause of death among women, globally.[21] This illustrates a systematic harm to women that is not the consequence of discrimination based on *observed* or *imagined* bodily features, but of intrinsic biological properties, such as different cardiovascular systems, combined with taking the male human body as the default human body.

In this section, I have argued that having, or introducing, phrases or concepts for (at least some) social kinds does not suffice to bring those kinds into existence: what has been named "witch trials" are not trials of witches. People labelled "enemies of the people" were victims of ideological political persecutions, but not traitors to their motherland. This is one way of refuting generic claims to the effect that (ideological) concepts shape, cause and constitute, social reality. I have also argued that this is not a problem for Haslanger's original proposal which *introduces* new concepts for the purposes of social theory. Against Simion's objection, I have offered reasons to think that Haslanger's analysis of gender terms does exceptionally well for the aims of the theory. For the same reason, the insistence on the folk concept or on the dictionary definition does not address the theoretical motivation behind her ameliorative analysis. I have further pointed towards empirical evidence about the systemic discrimination of women, which warrants the pursuit of the ameliorative project. Haslanger's proposed gender concept thus plausibly represents, but neither creates shape, women as a subordinate social class.

6.4 Concepts Are Not Necessary for "Shaping" Reality

As argued in the previous section, there are good reasons to think that Haslanger's analysis of gender as a social class both meets the aims of materialist feminist social theory and is supported by empirical evidence. In this section, I argue that concepts for social kinds are also not necessary for "shaping" reality.

In their responses to Simion, Podosky and McKenna make some arguable claims about the role of concepts and ideology in relation to social reality. Podosky said,

---

[21] See also https://www.thelancet.com/journals/lancet/article/PIIS0140-6736(19)30510-0/fulltext

> *[I]deological concepts are accurate because they constitute social reality*... Despite the accuracy of ideological concepts, we should not preserve them if they perpetuate oppression, no matter the epistemic cost. (Podosky, 2018, p. 2. My emphasis)

Podosky then says that a "revised concept has the capacity to causally influence the world and can therefore make itself representationally accurate" (Podosky, 2018, p. 2). Now, I don't mean to misrepresent Podosky's view. On one reading, he appears to defend primarily a causal account. Ideological concepts would merely have causal effects on social reality. But, although presumably true, it is a narrow and trivial understanding of the role of ideological beliefs, or of beliefs more generally. But it is inconsistent with the rest of his claim, "that ideological concepts constitute social reality" since beliefs are not veridical by default. First, it is trivial because it amounts to the claim that people's beliefs are explanatory of their actions and thereby have causal effects on the world. This is compatible with *false* ideological beliefs that do not "make themselves representationally accurate". The fact that some believed that others were witches had causal effects on the world: those accused were executed. It may be that those ideological beliefs are false *post hoc* rationalizations of the real motives and purposes of the accusers. Those *post hoc* rationalizations have effects, but they do not thereby become *accurate*, i.e., true representations of reality.

On a more interesting reading, ideological concepts would contribute to the constitution of social kinds. However, the counterexamples provided are also problems for this reading. Nonetheless, the objections I give here do not preclude the possibility of a more interesting story about how social kind concepts for possible social kinds can come to be representations of actual social kinds or help to bring such kinds into existence. Henceforth, I shall read Podosky as saying that new ideological concepts cause and can constitute a new social kind, which I think is also in accordance with the quote above.

McKenna similarly argues that "social kind concepts—have the primary function of shaping the world." (McKenna, 2018, p. 336) But what does it mean to say that a social kind concept has such a function? The ameliorated concept $woman_{SC}$ is not one, we hope, which will *create* new social positions of oppression for women. First, because those positions are already occupied. Second, if those positions were not occupied by females, we should *not* seek to create them!

To support his claim, McKenna alludes to *husband* and *wife* as social kind concepts. This suggests that he has the *constitutive* sense of "shaping reality" in mind, not the causal one. When a man and a woman are declared married, they *become* husband and wife, if the ceremony takes place in the right social circumstances. The ceremony does not just *cause* them to be married (although it also does that). However, for there to be the social kind *marriage* is not for there to be a concept for *marriage* and for people to be declared married *tout court*. There must be conventions that are adopted and enforced, with a purpose, which in turn satisfies an ongoing need. The mere grasp, or possession, of the concept itself does not constitute the contractual social reality of marriage.

Carole Pateman, a materialist feminist, described the marriage contract as a means to perpetuate the subordination of women, presented in a way that obscures its real nature (Pateman, 1988). The ordinary concepts of *husband* and *wife* do not disclose "the real nature" of marriage (if Pateman was right):

> Contract is the specifically modern means of creating relationships of subordination, but because civil subordination originates in contract, it is presented as freedom. (Pateman, 1988, p. 118)

Pateman's critical assessment of the sexual contract echoes Stuart Mill's reflection that "Marriage is the primitive state of slavery lasting on" (Mill, 1970, p. 130). If Pateman was right, the marriage contract would have served men's need to control access to women, while masquerading its real nature as a seeming embodiment of the liberal value of freedom. Thus, materialist feminists hoped to achieve a socially more just world by advocating for the abolishment of marriage.

Alas, *marriage* is typically offered as paradigm of conceptual amelioration. The focus is typically on jurisdictions that have changed the law to recognize same-sex marriages. Yet, the example is often under described. It was the change in the social conventions, and greater tolerance towards homosexuals, that allowed for the legal recognition of same-sex marriage. Discussions of the amelioration of *marriage* tend to also leave out other important revisions in the law. For example, many countries eradicated subordinating husband-wife contracts by, e.g., recognizing and criminalizing marital rape, by allowing married women to have financial and personal autonomy, the right to divorce, to keep custody of children, etc. The idea that it was the *concept* that shaped the new social reality of marriage is thus insufficiently motivated. Social and legal reality changed in good measure through the efforts of feminist and gay rights campaigners.

Now, we could wonder whether the concept $woman_{HF}$ is an ideological concept that we should not have. The question that materialist feminists struggled to answer was precisely, in Atkinson's words: "why, if women were not naturally inferior, our oppression had continued fundamentally unchanged." Surely, it is not the fact that there is an externally individuated concept for human females that has kept women in systematic positions of subordination, historically and globally. The folk concept is not an ideological one. If anything, what keeps women oppressed is the exploitation of their sexual and reproductive labour, the alienation from their value as people, the violence inflicted on women, etc. It is not the pre-theoretical folk that causes, constitutes, or brings about our ongoing subordination.

Striving to answer something akin to Atkinson's question led Haslanger to introduce a distinct concept, $woman_{SC}$. But it also cannot be that the *social kind concept*, introduced for the aims of social theory, is what caused or constituted the social kind. The new concept $women_{SC}$ is a concept of materialist feminist theory, even if related to the folk concept $woman_{HF}$. It is meant as a tool, within the theory, to better understand the social disadvantage occupied by $women_{HF}$. The fact that we lacked it, that it was introduced only in recent decades, while women's subordination has existed "as far back as history can be traced", as De Beauvoir said, shows that the social kind was neither caused nor constituted by the concept. This should not be controversial. Many non-German speakers took pleasure in schadenfreude before acquiring the concept for the feeling of pleasure at others' misfortune. Men have been mansplaining probably for as long as women have been discriminated, but the term and concept for the practice only recently came into use.

In conclusion, the problem with some conceptions of ameliorative projects is not merely that it is unclear whether social kind concepts have the function of "shaping" social reality. The problem is also that although people's beliefs and conceptions can have effects in the social world, these are not necessarily "accurate" or "veridical". These problems do not however affect the ameliorations that can be advanced to better *understand* social reality as it is. This, as I've argued in this paper, was the primary goal in Haslanger's 2000 paper on gender and race.

There may be another way to interpret and defend the claim that social kind concepts shape social reality. In this paper, I have argued against one version of this view—that a social kind concept can be revised or changed at will, and that so doing creates the corresponding social kind. I have also tried to rehabilitate the original motive behind materialist feminism: to look for the best theoretical tools to better understand the social world, as it impacts on women's lives. A different question, not addressed here, is whether it is necessary to devise new senses for old words to achieve that theoretical goal.[22]

References

Andler, M. (2017). Gender identity and exclusion: A reply to Jenkins. *Ethics, 127*(4), 883–895. https://doi.org/10.1086/691583

Atkinson, T. (2014). The descent from radical feminism to postmodernism. Presentation on the panel "How to defang a movement: Replacing the political with the personal" at the conference *A revolutionary moment: Women's liberation in the late 1960s and the early 1970s* organized by the Women's, Gender & Sexuality Studies Program, Boston University, March 27–29 2014.

Bach, T. (2012). Gender is a natural kind with a historical essence. *Ethics, 122*(2), 231–272. https://doi.org/10.1086/663232

Bach, T. (2016). Social Categories are Natural Kinds, not Objective Types (and Why it Matters Politically). *Journal of Social Ontology, 2*(2):177–201.

Barnes, E. (2020). Gender and gender terms. *Noûs, 54*(3), 704–730. https://doi.org/10.1111/nous.12279

[22] I'd like to thank the editors of this collection, Manuel Gustavo Isaac, Steffen Koch, and Kevin Scharp, and an anonymous reviewer. I'd also like to thank the audiences at the events where I presented earlier versions of the paper for their very helpful comments and suggestions: at the *Conceptual Engineering* Colloquium at GAP11, at the Humboldt University, in 2022; at the *LEME—Linguagem, Epistemologia e Metafísica* seminar of the Universidade Federal do Rio de Janeiro in 2020, and at the ARCHE Conceptual Engineering Seminar also in 2020, in particular to Célia Teixeira, Rosi Morokawa, Sally Haslanger, Guido Löhr, Mona Simion, Paul-Mikhail Catapang Podosky, and Sarah Sawyer. Special thanks to Duccio Calosi, with whom I've spent many hours discussing some of the fundamental problems related to conceptual engineering. Work towards this paper was financially supported by grant CEX2021-001169-M funded by MCIN/AEI/10.13039/501100011033, by the project PID2023-150569NB-I00 of the Spanish Ministry of Science and Innovation, and by 2021-SGR-00276 for the Consolidated Research Group LOGOS, by the Generalitat de Catalunya.

Bogardus, T. (2020). Some internal problems with revisionary gender concepts. *Philosophia, 48*(1), 55–75. https://doi.org/10.1007/s11406-019-00107-2

Byrne, A. (2020). Are women adult human females? *Philosophical Studies, 177*(12), 3783–3803. https://doi.org/10.1007/s11098-019-01408-8

Cantor, N. (1993). *The civilization of the middle ages: A completely revised and expanded edition of medieval history, the life and death of a civilization*. Harper Collins.

Constitution of India, article 15. Retrieved November 2023, from https://www.constitutionofindia.net/articles/article-15-prohibition-of-discrimination-on-grounds-of-religion-race-caste-sex-or-place-of-birth/

Criado Perez, C. (2019). *Invisible Women: Data Bias in a World Designed for Men*. Abrams Press.

de Beauvoir, S. (1949/2010) *The Second Sex*, Constance Borde and Sheila Malovany-Chevallier (trs.), Random House.

Figes, O. (2007). *The whisperers: Private life in Stalin's Russia*. Penguin Books.

Glass, B. L., & Stolee, M. K. (1987). Family law in soviet Russia, 1917–1945. *Journal of Marriage and Family, 49*(4), 893–902. https://doi.org/10.2307/351982

Goddard, C. (2002). Lexico-semantic universals: A critical overview. *Linguistic Typology, 5*, 1–65. https://doi.org/10.1515/lity.5.1.1

Haslanger, S. (2000). Gender and race: (What) are they? (What) do we want them to be? *Noûs, 34*(1), 31–55. https://doi.org/10.1111/0029-4624.00201

Haslanger, S. (2020). How Not to Change the Subject. In Teresa Marques & Åsa Wikforss (editors), *Shifting Concepts: The Philosophy and Psychology of Conceptual Variability*. Oxford: Oxford University Press.

Hayden, B., Deal, M., Cannon, A., & Casey, J. (1986). Ecological determinants of women's status among hunter/gatherers. *Human Evolution, 1*(5), 449–473. https://doi.org/10.1007/BF02436536

Isaac, M. G. (2024). Post-truth conceptual engineering. *Inquiry, 67*(1), 199–214. https://doi.org/10.1080/0020174X.2021.1887758

Jenkins, K. (2016). Amelioration and inclusion: Gender identity and the concept of woman. *Ethics, 126*(2), 394–421. https://doi.org/10.1086/683535

Kasrnovsky, H. (2021). Pre-agriculture gender relations seem bad. *Cold Takes* blog/podcast. Retrieved January 2022, from https://www.cold-takes.com/hunter-gatherer-gender-relations-seem-bad/

Kitsik, E. (2023). Epistemic paternalism via conceptual engineering. *Journal of the American Philosophical Association, 9*(4), 616–635. https://doi.org/10.1017/apa.2022.22

Laskowski, N. G. (2020). Moral constraints on gender concepts. *Ethical Theory and Moral Practice, 23*(1), 39–51. https://doi.org/10.1007/s10677-020-10060-9

Lawford-Smith, H. (2022). *Gender-critical feminism*. Oxford University Press. https://global.oup.com/academic/product/gender-critical-feminism-9780198863885?cc=es&lang=en&#

Lerner, G. (1986) *The Creation of the Patriarchy*. Oxford University Press.

Liberman, A. (2011). Were ancient 'wives' women? In *OUPBlog – Oxford University Press's insights for the thinking world*. https://blog.oup.com/2011/10/wife/. Accessed Nov 2023.

Machery, E. & Faucher, L.(2020). The Folk Concept of Race. In Teresa Marques & Åsa Wikforss (editors), *Shifting Concepts: The Philosophy and Psychology of Conceptual Variability*. Oxford: Oxford University Press.

Malt, B. (2020). Mapping thoughts to words: Cross-language difference, learning, and communication. In T. Marques & Å. Wikforss (Eds.), *Shifting concepts: The philosophy and psychology of conceptual variability* (pp. 17–33). Oxford University Press. https://doi.org/10.1093/oso/9780198803331.003.0002

Marques, T. (2017). The relevance of causal social construction. *Journal of Social Ontology, 3*(1), 1–25. https://doi.org/10.1515/jso-2016-0018

Marques, T. (2020). Amelioration vs. perversion. In T. Marques & Å. Wikforss (Eds.), *Shifting concepts: The philosophy and psychology of conceptual variability* (pp. 260–284). Oxford University Press. https://doi.org/10.1093/oso/9780198803331.003.0014

Marques, T., & Wikforss, Å. (2020). *Shifting concepts: The philosophy and psychology of conceptual variability*. Oxford University Press. https://doi.org/10.1093/oso/9780198803331.001.0001

Marx, K. (1975). *Early Writings*, ed. L. Colletti. Penguin, Harmondsworth.

Marx, K., & Engels, F. (1975). *Selected correspondence*. Progress.

McKenna, R. (2018). No epistemic trouble for engineering 'woman'. *Logos and Episteme, 9*(3), 335–342. https://doi.org/10.5840/logos-episteme20189325

Meyerson, D. (1991). *False consciousness*. Clarendon Press.

Mikkola, M. (2009). Gender concepts and intuitions. *Canadian Journal of Philosophy, 39*(4), 559–583. https://doi.org/10.1353/cjp.0.0060

Mill, J. S. (1970). The subjection of women. In A. S. Rossi (Ed.), *Essays on sex equality*. University of Chicago Press.

Mills, C. W. (2014). Notes from the resistance: Some comments on Sally Haslanger's *Resisting Reality*. *Philosophical Studies, 171*(1), 1–13. https://doi.org/10.1007/S11098-013-0249-9

Pateman, C. (1988). *The sexual contract*. Polity Press.

Podosky, P. (2018). Ideology and normativity: Constraints on conceptual engineering. *Inquiry: An Interdisciplinary Journal of Philosophy*, 1–15. https://doi.org/10.1080/0020174X.2018.1562374

Podosky, P. (2022). Don't count truth out just yet: A response to Isaac. *Inquiry: An Interdisciplinary Journal of Philosophy*, 1–12. https://doi.org/10.1080/0020174X.2022.2075922

Prasso, S. T. (2001). The riel value of money: How the world's only attempt to abolish money has hindered Cambodia's economic development. *Asia-Pacific Issues. Analysis from the East-West Center*, (49). Retrieved November 2023, from https://www.eastwestcenter.org/publications/riel-value-money-how-worlds-only-attempt-abolish-money-has-hindered-cambodias-economic-

Putnam, H. (1975). The meaning of 'meaning'. In *Philosophical papers* (Vol. 2, pp. 215–271). Cambridge University Press.

Queloz, M., & Bieber, M. (2021). Conceptual engineering and the politics of implementation. *Pacific Philosophical Quarterly*, (3), 670–691. https://doi.org/10.1111/papq.12394

Ringe, D. (1992). On calculating the factor of chance in language comparison. *Transactions of the American Philosophical Society, New Series, 82*(1), 1–110. https://doi.org/10.2307/1006563

Rivera, C., Hsu, Y., Pavez Esbry, F., & Dugarova, E. (2020, July 13). *A dashboard overview: Gender inequality and the COVID-19 crisis*. United Nations Development Programme, Human Development Reports. Retrieved October 2020, from http://hdr.undp.org/en/content/dashboard-overview-gender-inequality-and-covid-19-crisis

Saul, J. (2006). Gender and race. *Aristotelian Society Supplementary Volume, 80*(1), 119–143. https://doi.org/10.1111/j.1467-8349.2006.00140.x

Sawyer, S. (2020). Truth and objectivity in conceptual engineering. *Inquiry, 63*(9–10), 1001–1022. https://doi.org/10.1080/0020174X.2020.1805708

Schiff, S. (2015). *The Witches: Salem, 1692*. Little, Brown and Company.

Shields, M. (2021). Conceptual domination. *Synthese, 199*(5–6), 15043–15067. https://doi.org/10.1007/s11229-021-03454-4

Simion, M. (2018a). The 'should' in conceptual engineering. *Inquiry: An Interdisciplinary Journal of Philosophy, 61*(8), 914–928. https://doi.org/10.1080/0020174X.2017.1392894

Simion, M. (2018b). Epistemic trouble for engineering 'woman'. *Logos and Episteme, 9*(1), 91–98. https://doi.org/10.5840/logos-episteme2018916

Simion, M., & Kelp, C. (2019). Conceptual innovation, function first. *Noûs, 54*(4), 985–1002. https://doi.org/10.1111/nous.12302

Wierzbicka, A. (1996). *Semantics: Primes and universals*. Oxford University Press. https://doi.org/10.1093/oso/9780198700029.001.0001

Wolff, J. (1992). Playthings of alien forces: Karl Marx and the rejection of the market economy. *Cogito, 6*(1), 35–41. https://doi.org/10.5840/cogito19926117

Chapter 7
Engineering Ideologically Defective Concepts

Mari Mikkola

Abstract Conceptual engineering typically aims to fix some conceptual defects. Mona Simion, however, argues that it is permissible to engineer non-defective concepts too, but this mustn't come with an epistemic loss. This paper holds that some conceptual resources should be fixed though because of serious ideological defects involved and it isn't possible to engineer such concepts without an epistemic loss; still, this is the normative ground *to* engineer. In this spirit, Paul-Mikhail Catapang Podosky argues for an expanded version of Simion's view to make engineering ideologically defective concepts permissible. Although Podosky's critique of Simion is compelling, this paper critiques Podosky's expanded version too: Podosky's understanding of how ideological concepts work also generates prima facie prohibitions to engineer when engineering looks highly desirable. The paper thus proceeds to offer an alternative diagnosis of ideological defects. It subsequently raises another, less normative, issue about implementation: it is extremely difficult, if not impossible, to merely 'tweak' ideological concepts. The discussion hence offers grounds to be skeptical about the feasibility of implementing new conceptual resources. It also urges modesty about how emancipatory 'our' attempts at conceptual innovation can be without this being enabled by alterations in background social conditions.

M. Mikkola (✉)
University of Amsterdam, Amsterdam, Netherlands
e-mail: m.mikkola@uva.nl

M. G. Isaac et al. (eds.), *New Perspectives on Conceptual Engineering - Volume 2*, Synthese Library 514, https://doi.org/10.1007/978-3-031-98524-9_7

7.1 Introduction[1]

One major motivation for engaging in conceptual engineering is that some apparent ways in which we think and talk about reality are *defective*. The kinds of defects that the engineer may wish to correct and fix include:

> cognitive defects (that undermine our ability to reason properly), moral or political defects (that undermine moral or political values of various sorts), theoretical defects (that undermine progress within some theoretical field), or semantic defects (where the semantic value is incoherent, incomplete, or missing). (Cappelen & Plunkett, 2020, p. 3; see also Cappelen, 2018)

For instance, take the concept of *family*: traditionally conceived, its content encompasses a mother and a father, who are bound together by a heterosexual marriage contract, and have biologically related offspring. But this conception is defective in various way. Morally and politically, it generates failures to recognize less traditional arrangements in our social milieu: same-sex unions, adoptive parents, parenting one's partner's biological children, non-married couples, childfree unions, and so on. It seemingly involves cognitive defects in suggesting that a merely contingent social formation is a necessary one. Having just this conception in one's theoretical apparatus is likely to result in impoverished theoretical understandings of contemporary social life. And such theoretical defects may generate semantic defects via conceptual eclipsing (Haslanger & Ásta, 2018): 'our' conceptual resources allow us to spot certain phenomena, while eclipsing others. Subsequently, it looks as if the conceptual engineer is first and foremost in the business of identifying some defects in need of correction and then "changing existing concepts and devising new ones" accordingly (Braddon-Mitchell, 2020, p. 79; see also Cappelen, 2018; Chalmers, 2020). As identified by Isaac et al. (2022), engineering involves various components. First, the assessment component evaluates an existing conceptual or representational device by considering (for example) how well it fulfils its function (whatever that may be)—roughly, it involves identifying how some conceptual device is defective. Second, the improvement component aims for some conceptual innovation by bettering the functional quality of the device under consideration—seeking to fix the deficiencies identified. Third, implementation component is about how we can ensure that the engineered and improved conceptual resources will be taken up by speakers.[2] In other words, the assessment component guides the improvement/ innovation component, which in turn affects the implementation component.

[1] I have presented earlier versions of this paper at CEU (Vienna) and the Conceptual Engineering Online Seminar organized by the editors of this volume. I am deeply grateful to those present for their insightful queries and challenges. Furthermore, I am hugely indebted to the editors of this volume and anonymous commentators for their extensive comments and difficult questions on an earlier version of the paper.

[2] Isaac et al. (2022) identify a fourth, descriptive component too. But I won't consider it here because it does not bear on my subsequent discussion.

Despite the common focus on fixing defective language, Mona Simion (2018) argues that it is permissible to engineer non-defective concepts too: engineering need not involve the idea that some pre-existing defect must be fixed, but the idea that we can improve our linguistic resources and repertoire even when they are (bluntly put) good enough. Still, Simion holds that this should not be unrestricted and there needs to be normative constraints on when it is legitimate to engineer an otherwise non-defective concept: this is legitimate when we have an all-things-considered reason to engineer and "in the process of revising the concept, we do not end up with less knowledge about the world" (Simion, 2018, p. 924). Engineering must not come with an epistemic loss—a criterion that Manuel Gustavo Isaac (2021a) calls the 'alethic criterion'.

I do not dispute that we can have reasons to engage in conceptual improvement in the sense of continual maintenance, instead of just focusing on correcting serious defects. There are, nonetheless, some conceptual resources that many hold we should—in fact, must—fix because the defects involved are too serious. These are ones that I term 'ideologically defective concepts.' I hold that it isn't possible to engineer such concepts without an epistemic loss—but contra Simion this isn't something that should count against engineering; rather, it is precisely the normative ground to engineer. I am not alone in proposing this line of thought. Paul-Mikhail Catapang Podosky (2018) argues for an expanded version of Simion's alethic criterion precisely since Simion's constraint would make engineering ideologically defective concepts impermissible (for an alternative discussion of Simion similar to Podosky in spirit though not in letter, see Valero, 2022). Podosky holds that it is legitimate to engineer (or, in his words, ameliorate) ideological concepts "where the revised concept has the capacity to causally influence the world, and can therefore make itself representationally accurate." (2018, p. 2) This revised constraint, however, comes with a feasibility criterion:

> The claim that a concept should be engineered is right if its desired causal influence on social reality is feasible; and the claim that a concept should be engineered is wrong if its desired causal influence on social reality is infeasible. Importantly, the notion of feasibility mentioned here is epistemic: It is about the feasibility of representational accuracy. (Podosky, 2018, p. 13)

I think that Podosky is right about ways in which Simion's normative constraint falls short. However, his feasibility constraint too leaves much to be desired when thinking about ideologically defective concepts. Or so I will argue here. The crux of my case hinges on thinking about what is ideologically defective about certain concepts. Podosky's understanding of how ideological concepts work is not apt to spell out what it means for conceptual defects to hinge on ideology—in other words, I disagree with his assessment of the defects to be fixed. This in turn (I hold) misconceives the needed conceptual improvements and misdirects how innovations can be implemented. Advancing an alternative diagnosis of how ideologically defective concepts work, demonstrates that Podosky's reformulation of Simion's normative constraint also falls short: it generates prima facie prohibitions to engineer when it seems that engineering is precisely what we ought to engage is. That is, when applied to concepts putatively defective in ideological ways, Podosky's feasibility criterion coupled with his understanding of ideology results in quietism about improving such concepts and about fixing ideological defects. This undercuts the grounds to engineer ideologically defective concepts, and hence yields the wrong normative constraint to fix the permissibility to engineer.[3]

My alternative diagnosis of the defects, however, raises another, less normative, issue about implementation. It demonstrates that it is indeed extremely difficult, if not impossible, to engineer ideological concepts even when we have good normative grounds to do so. Thinking more carefully about ideological defectiveness suggests that de facto the prospects of engineering away ideological defects are extremely poor—at least, in David Chalmers's (2020) sense of re-engineering them. Still, I go onto consider whether an alternative account of engineering is more promising, namely one that takes conceptual engineering to encompass conceptual ethics. On some accounts, paradigm engineering projects are "exercises in conceptual innovation, guided by some conceptual ethics, with an eye (and likely some effort put toward) effecting actual conceptual change"—in other words, toward conceptual implementation (Burgess & Plunkett, 2020, p. 286). So, the engineer might engage in *de novo* engineering, as Chalmers puts it, together with conceptual ethics: devising new conceptual resources and "making normative judgments as to whether they are fit for purpose" (Braddon-Mitchell, 2020, p. 79) relative to one another and/or some older resources. Given my assessment of ideologically defectively concepts, I hold, the implementation stage of this sort of conceptual engineering will be tricky as well: conceptual innovation rather looks to piggy bag structural changes in background social conditions, rather than being the driving force behind such changes, which puts pressure on whether engineering is feasible in the sense of implementation.

I will start by sketching out Simion and Podosky's positions (Sect. 7.2). Then I will consider what is ideological about ideologically defective concepts. This requires taking slight detours to discussions about non-ideal theory (Sect. 7.3), ideology critique (Sect. 7.4), and constitutive foreclosure (Sect. 7.5) that fall outside of typical discussions around conceptual engineering. They do, nevertheless, serve important argumentative ends in offering an alternative understanding of what the ideological nature of ideologically defective concepts amounts to. I will return to Podosky's feasibility constraint in (Sect. 7.6) and conclude with a discussion of the possible benefits of *de novo* engineering coupled with conceptual ethics (Sect. 7.7).

[3] Note that the objection I am making against Podosky pertains to ideologically defective concepts, which are ones that Podosky too (as far as I can tell) thinks ought to be improved. My objection isn't about engineering concepts defective in some other manner. There can clearly be reasons not to engineer concepts that are defective in some non-ideological ways if the feasibility criterion isn't met. But I am here not focusing on engineering tout court, but rather on engineering concepts that look to be defective in a specific way. Podosky's view (to me) is precisely that we should revise Simion's normative criterion to allow for engineering ideologically defective concepts. Hence, it looks like Podosky should want to resist quietism about engineering ideologically defective concepts. But if his feasibility criterion results in such quietism, this looks to be a serious problem for the view he puts forward.

7.2 Alethic Criterion

Let me start by saying more about Simion and Podosky's views. Simion holds that: "Concepts, just like belief, are representational devices, their function is an epistemic one: to represent the world…Concepts will be good concepts qua concepts when they are epistemically good." (2018, p. 923) This view straightforwardly supports the alethic criterion: that conceptual revisions should not lead to epistemic losses. However, as noted, Podosky holds that when engineering politically and socially dysfunctional concepts—or ideological concepts, for short—epistemic loss is often inevitable *and* desirable. We should then resist Simion's constraint.[4] Podosky has a particular understanding of ideology, which Isaac (2021a) calls Haslangerian-like constructionist analysis. For Podosky, "*ideology* serves to create social reality; it makes certain facts true. And ideology consists of concepts. Hence, ideological concepts are *accurate* because they constitute social reality." (2018, p. 2) Podosky takes ideologies to be "representations of social reality that serve to govern the practices of agents embedded in social structure" (2018, p. 4). However, since ideology is oppressive, we should not aim to preserve the kind of knowledge that it generates. And so, knowledge preservation does not yield the correct normative constraint for engineering. If ideological concepts can literally make social reality be some way, engineering them will result in an epistemic loss and this would hence render engineering impermissible on Simion's alethic criterion—which looks to be the wrong result.

Instead, for Podosky, legitimate engineering can result in an epistemic loss but *only* in a temporary one. Certain social concepts have the causal capacity to make themselves accurate if taken up by a community of speakers (Podosky, 2018, p. 9). Perhaps there is a temporary epistemic loss in the process of a community taking up the concept; but ultimately the social world arranges itself to fit the revised concept. And so, "if the newly revised concept is widely accommodated [i.e. has made itself accurate by self-reinforcement], and has influenced social reality, there will be no epistemic deficit; there will be no overall epistemic loss" (Podosky, 2018, p. 11). This nevertheless requires that the newly engineered concepts *feasibly* have the causal capacity to make themselves accurate. If they do, we have grounds to

[4] It is worth noting that Simion does not apparently think such an epistemic loss follows on her account. She writes: "If our language is constitutive of social reality, conceptual engineering will amount to a form of social engineering … since as soon as one engineers a social kind concept, reality follows suit, it is unlikely to ever get epistemic deficiencies. As such, by [my criterion], any type and extent of improvement, not matter how marginal or limited, justifies an ameliorative project" (2018, p. 926). If Simion is right about reality following suit, I agree that engineering is always permissible on her account. However, I think that reality does not follow suit. Or rather: there is distinction between social kind concepts and ideologically defective (social kind) concepts. If engineering social kind concepts were so easy, we would have no issues with such concepts being somehow ideological or defective. But the challenge I am considering derives precisely from some (perhaps most) social kind concepts being sticky-ly defective, where reality does not follow suit. With them in mind, it looks like the alethic criterion speaks against engineering when it looks like engineering is precisely what we should engage in.

engineer. If they do not, we have grounds not to engineer. To have such causal capacity is akin to a disposition "to become accurate under the right conditions" (Podosky, 2018, p. 10, fn 18). That is, to have the capacity "to make itself accurate; the concept *could*, in principle, accurately represent how things are without the help of revision" (Podosky, 2018, p. 10).

To me, Podosky's critique of Simion is compelling. But I do not find his own alternative or the feasibility criterion so. For a start, the analysis of ideology he offers is underdeveloped and does not offer a compelling account of what is *dysfunctional* about some representations. Not all representations of social reality are ideologically defective or defective in any manner. Merely adding that these representations must be oppressive to be ideological does not yet offer an informative analysis of such supposed ideological nature. More importantly though, Podosky's account of how concepts have the causal capacity to make themselves accurate looks to have some internal tensions. For ideological concepts to have the causal capacity to make reality conform to them and for a community to take up engineered concepts, the right background social conditions must obtain. How would these conditions come to be? I take it that conceptual engineers, including Podosky, think that we can bring about structural social change by engineering away ideological defects in our conceptual repertoire and representational devices. In other words, once the improved conceptual resources have been taken up by speakers, this will have real world effects as 'our' social structures will arrange themselves to fit the improved conceptual recourses. But for Podosky, engineering is permissible only if the engineered concepts could feasibly have the causal capacity to be taken up and so affect social reality. To engineer, it must be possible to causally affect social reality. However, since this is to be understood dispositionally, social reality must already be such that engineering will be feasible: the right social conditions must already obtain for engineering to have the causal capacity to be taken up by a community of speakers. Bluntly, the disposition "having the causal capacity to become representationally accurate" encompasses the right stimulus conditions: that favorable background conditions obtain. Without these background conditions, there is no such causal capacity. Engineering won't be feasible and hence, we ought not to engineer. What is doing the heavy lifting then is whether the background social conditions are functional or dysfunctional—in the latter case, the feasibility criterion will make engineering impermissible when it looks like engineering is precisely what we ought to engage in. If the social conditions are of the right kind though, then engineering is neither here nor there. In short, engineering looks to be either impossible or emancipatorily irrelevant.

This is in the spirit of Isaac's (2021a) critique of the alethic criterion. For him, ideology is not about truth, but about power. The main issue for conceptual engineering hence is conceptual efficacy, rather than truth-trackingness: how well do some conceptual resources "fulfil their function of ultimately enabling the agent to execute the cognitive task they are used for" (Isaac, 2021a, p. 10). Elsewhere Podosky (2022) replies to Isaac, but (I contend) in a manner that does not dispel my concerns above. According to Podosky, Isaac's critique misses its target because for Podosky truth *is* power. To quote Podosky:

> this can be seen when we look beyond the idea that concepts *shape* social reality. Shaping is just one means by which power is exercised or instantiated. But it is not the only means. Power is also evident and present in the *maintenance* of the status quo. Ideological concepts do not just come about, but rather they are robust, sticky, and hard to dislodge. And this stabilizes unjust social practices, and stifles progress. (2022, p. 5)

But now Podosky's feasibility criterion becomes more puzzling still. It is legitimate to engineer if engineering has the causal capacity to affect social reality. What undergirds this causal capacity, then, is not merely that the "right" background conditions obtain but also the amount of social power the would-be engineers have. This may de facto be the case when thinking about real world implementation of engineered concepts. Nevertheless, Podosky is aiming to offer a normative constraint: when *should* we engineer. And so, claiming that truth is power looks to stifle the project of conceptual engineering even more if taken as the normative limit and constraint on engineering. Engineering will only be permissible (in being feasible) if the would-be engineers have enough social power to shape social reality and to dislodge those conceptual resources in need of revision. But if ideologically defective concepts are robust and sticky, where those who uphold such concepts have the social power to keep in place the status quo, emancipatory engineering turns out to be impermissible because it won't be feasible. With his reply to Isaac, Podosky ends up undermining his position in a slightly different manner: engineering will be permissible (in being feasibly) only after emancipatory-engineers have social power to affect social reality. It seems impossible though for ideological concepts to be dislodged through conceptual engineering, unless those doing the engineering *already* have sufficient social power to effect change. Once again, either engineering looks to be paralyzed unless it is enabled by background social structures and conditions; or engineering just looks irrelevant since the emancipatory heavy lifting is taking place through improvements to background structural conditions.

7.3 Non-Ideal Theory and Ideology

My contention is that we need an alternative account of the ideological nature of ideologically defective concepts to think about how to engineer away those defects. To address such nature, I will take a short and non-comprehensive detour to considering ideal/non-ideal theory in political philosophy. This detour is not random or idiosyncratic. Non-ideal approaches in political theory seemingly share much with conceptual engineering. They hold that there is something deeply defective and ideological about the conceptual resources and repertoire of (so-called) ideal theory; hence, it is instructive to look at how this apparent ideological defectiveness is conceptualized. To do so, consider (what I take to be) the central tenets of ideal theory according to one historically leading theorist, John Rawls. According to him, an ideal theory of justice "develops the conception of a perfectly just basic structure and the corresponding duties and obligations of persons", (Rawls, 1971, pp. 245–246). The subsequent ideal principles of justice that Rawls develops and articulates are meant to cover societies where "everyone accepts and knows that the others accept the same principles of justice" and "the basic social institutions generally satisfy and are generally known to satisfy these principles" (Rawls, 1971, p. 5). Moreover, Rawls assumes social agents' strict compliance with these principles of justice, hence aiming to articulate the ideal—perfect—conception of a just basic social structure (to put it bluntly).

The details of Rawls's principles are not important for my purposes. Rather, what matters are two methodological theses to which his view is wedded. First, what I call 'the normative priority thesis': that we need to articulate an ideal conception of justice before we can tackle non-ideal circumstances. The thought is that we need to know what perfect justice consists in first before we can examine extant social conditions to identify injustices through its lens. As Rawls writes: "Existing institutions are to be judged in the light of this conception [the ideal conception of a just society] and held to be unjust to the extent that they depart from it without sufficient reason" (1971, p. 246). Rawls takes this method to yield *better* theorizing in providing the basis for systematically grasping pressing real-world problems—but this is something that he explicit merely assumes (1971, p. 9). Second, Rawlsian ideal theory is committed to a methodological outlook, which I call 'the distancing thesis':

> the deeper the conflict, the higher the level of abstraction to which we must ascend to get a clear and uncluttered view of its roots … formulating idealized, which is to say abstract, conceptions of society and person … is essential to finding a reasonable political conception of justice. (Rawls, 1993, pp. 45–46)

Good theory building supposedly starts by distancing away from actual states of affairs, both current and historical. Thereby such theorizing relies on either possible, but non-actualized, or impossible *idealizations* about (for instance) the natures of citizens and political institutions.

Charles Mills prominently diagnosed that what goes wrong with this sort of theorizing is that it functions *ideologically* in a pejorative sense. Normative concepts like Rawlsian *justice* serve the needs of a small number of privileged persons thereby being obfuscating. The above two methodological outlooks generate normative concepts that involve "a distortional complex of ideas, values, norms, and beliefs that reflects the nonrepresentative interests and experiences of a small minority of the national population – middle-to-upper-class white males" (Mills, 2005, p. 172). For Mills, the distinguishing feature of what is objectionable about Rawlsian-type ideal theorizing is its reliance on idealizations to the exclusion and marginalization of actual circumstances, which distorts our analyses. In a similar vein, Onora O'Neill (1987) critiques idealization and defends abstraction, the latter of which she understands as leaving out, bracketing off, or selectively omitting some aspects of particular cases. Since all theorizing aims to be somehow general, abstraction is unavoidable in theory building. But idealizations are methodologically corrupting. For instance, post-Enlightenment descriptions of human agency posited super-human idealized versions of agency that only hypothetical agents could satisfy. The problem is not that these versions omitted something about true agency, but that they imported too much *false* information into our conception of human agency, like super-human cognitive capacities to be treated as standards for human action per se. Idealizing agency obfuscates 'our' understanding and account of human agency—the resulting conceptual resources, then, are defective and prima facie candidates for engineering.

7.4 False Versus Perniciously False Idealizations

The above debates in political philosophy suggest that what makes some conceptual resources ideologically defective is that they are obfuscating (in certain ways) in being idealizations. Now, one might immediately wonder whether the distinction between abstraction and idealization can so easily be drawn. Prima facie in idealizing (say) human agency in a manner that adds something false, we must conceivably first bracket off information about real agency. Hence, the idea that abstraction involves leaving out something while idealization involves adding something does not neatly distinguish the two since in leaving something out, we typically add something else and vice versa. This immediate reaction, however, points to something significant: as I will argue shortly, idealization isn't objectionable because it posits something false, but because it posits something *perniciously* false. The falsity in play must be obscuring and obfuscating in a manner that generates conceptual defects in need of engineering. This idea of pernicious falsity, I contend, is significant for analyzing the ideological nature of ideologically defective concepts.

To see the distinction more clearly, consider a conceptual repertoire that contains *frictionless plane*. Even though the concept is an idealization and therefore posits something false, it is not prima facie defective in the senses outlined earlier. It does not obviously generate moral and political defects; and, in fact, the distortedness of the concept is theoretically and cognitively beneficial. So, what distinguishes *agency* (involving superhuman and unachievable cognitive and volitional powers) from *frictionless plane*?

To help us understand what the ideological nature of ideologically defective concepts hinges on, Tommie Shelby's (2003) account is useful. His view draws on Raymond Geuss's (1981) classic discussion and provides a prominent alternative view of ideology to the one that Podosky relies on.[5] Most basically for Shelby, ideologies involve systems of beliefs that are appropriate objects of critique. Therefore, ideology-critique is to be directed at *beliefs*, which he takes to be the primary units of analysis. For instance, since it is racist beliefs that are disseminated, transmitted, and reinforced via racist practices and representations, Shelby maintains, this makes those beliefs more fundamental than practices and representations (2003, p. 158). Furthermore, ideologies involve a network of beliefs. They seemingly have some

[5] Shelby aims to offer an elucidation of racism as an ideology from a Marxian perspective. Insofar as this is deeply materialistic in the sense of being wedded to social class struggles, not all of his insights are relevant for my purposes here. Nonetheless, his characterization of what ideology consists in is helpful to understand the nature of ideological defectiveness.

compelling explanatory or justificatory power for those who embrace them; such systems of thought tend to be deeply entrenched and constitutive of 'common sense'; and they are "intimately tied to action and social practice" (Shelby, 2003, p. 159). The ideological character of such beliefs, then, is comprised of "a complex set of characteristics that has epistemic, genetic *and* functional dimensions." (Shelby, 2003, p. 164)

First, ideological systems of beliefs have some defective *epistemic* properties; for example, they lack "empirical support, consistency, logical validity, conceptual clarity" (Shelby, 2003, p. 164). Second, such systems are problematic due to their *origin*—for example, if their adoption and endorsement are grounded in the believer's desire to perpetuate some unjust social relations. Third, ideological systems of beliefs *function* to have some negative practical consequences for society as a whole or for some social group. For instance, they function to keep in place unjust and systematic relations of power and privilege. Shelby's complex ideology-making feature involves all three dimensions in the forms of ideological illusions, false consciousness, and social functionality.

Ideological illusions are in no sense unreal, but rather involve cognitive defects leading to cognitive errors without typically involving straightforward falsity. Instead, they involve "distorted, biased, or misleading representations of reality", and errors such as "inconsistency, oversimplification, exaggeration, half-truth, equivocation, circularity, neglect of pertinent facts, false dichotomy, obfuscation, misuse of 'authoritative' sources, hasty generalization" (Shelby, 2003, p. 166). However, ideological illusions often go unnoticed or unrecognized by social agents in that ideological beliefs are regularly confirmed by casual observation. The resulting false consciousness, then, is somewhat of a misnomer: it is not about straightforwardly having false beliefs content-wise (though it might be). Rather, such consciousness is about the *way* in which beliefs are held (Shelby, 2003, p. 170)—for instance, when one is self-deceiving about the motives on the basis of which one is holding a belief. In committing cognitive errors of the kind noted, a cognizant may accept something due to the influence of some noncognitive motives, rather than have some solid epistemic ground for the belief. And they may be ignorant of this in thinking that they have based their belief on cognitive (rather than noncognitive) grounds—just think of confirmation bias. Holding a belief with false consciousness, then, involves being under "the unconscious influence of noncognitive motives" (Shelby, 2003, p. 170), which is irrational. Finally, ideological systems of belief function to legitimate and reify unjust social relations through illusion and misrepresentation: "Reifying ideologies stabilize oppressive social relations by creating the illusion that these relations (or their causes), which are actually the product of historically contingent human action or convention, are 'natural' and, thus, ineradicable, unavoidable, and unalterable." (Shelby, 2003, p. 177) Whatever way ideologies do their "dirty work" (as Shelby puts it), they all have the result of preventing social agents from recognizing or appreciating that their actions are implicated in oppressive social relations (Shelby, 2003, p. 177).

Having elaborated these dimensions, let me address a possible immediate objection. One might hold that ideological systems of beliefs have little to do with conceptual content. That is, one might have perfectly good conceptual resources but come to hold many false beliefs because ideologies distort our doxastic functioning. Alternatively, one can have a defective conceptual repertoire for nonideological reasons. Following Shelby, one might hold, ideological defects work at a doxastic, rather than a conceptual, level: there may be nothing defective about the categories that my conceptual resources pick out. Instead, I have come to form some false beliefs or made some infelicitous inferences about the members of those categories on ideological grounds. I do not wish to engage in a lengthy discussion of how beliefs and concepts are connected, but I do not think that my focus on the conceptual level is misguided. Or rather, I do not think that we can so neatly separate the doxastic and conceptual levels when dealing with ideological conceptual defects. Think about having the traditional conception of *family* in one's conceptual toolbox. This conception looks to be defective in that it generates conceptual eclipsing: it focuses on some phenomena, thereby ignoring others. This defect at the conceptual level plausibly leads to defects on the doxastic level, which then again reinforce the conceptual defects via cognitive errors that are motivated noncognitively and that function to create and maintain unjust social relations. It is precisely because ideological defects are *not* primarily about having false beliefs that they are so difficult to overcome: people whose conceptual repertoire is strongly wedded to the traditional notion of *family* precisely do not tend to revise that repertoire when presented with putative or obvious counterexamples. Rather, they tend to explain away such apparent challenges. Hence, ideological systems of belief are plausibly underpinned by conceptual defects of some kind. We might then say in terms of Isaac et al. (2022) that the *purpose* of engineering is to engage in ideology critique to foster social justice, while the *goal* is to tackle the conceptual repertoire that renders ideologies so sticky. Still, to reach our goal for the purpose given, we need to engage in assessment: to identify how some conceptual device is defective and, in this case, how it is ideologically defective. For this task, I submit, Shelby's complex ideology-making feature (ideological illusions, false consciousness, and social functionality) is instructive even though he isn't primarily targeting 'our' conceptual resources.

Shelby's more fleshed out conception of ideology allows us to examine how idealizations can be not only false, but perniciously false. Consider two idealizations: the traditional conception of *family* and *frictionless plane*. For idealizations to be perniciously false, I hold, they need to satisfy the negative ideology-defining complex. First, they must involve some cognitive defects where falsity alone isn't doing the work. Both *family* and *frictionless plane* involve distortions, misleading representations of reality, oversimplifications, and exaggerations. But the latter looks less obfuscating: it isn't the case that a conceptual repertoire including *frictionless plane* goes unnoticed or unrecognized in being simply confirmed by casual observation apparently as non-defective. Being a scientific postulate, casual observation does not come into play here. The case of *family* in a traditional sense is different though; it has much more of a 'common-sensical' character to it. This is precisely why *family* looks to be an extremely good candidate for engineering: its defects tend to go unnoticed or unrecognized in that casual observation often deems this conception to be fitting, even when strictly speaking empirically this isn't the case. (In fact, I suspect that many ordinary speakers take the traditional conception to be analytic. They may be willing to accept alternative arrangements, but these would not properly fall under *family*. Instead, they might fall under *household* or *living arrangements*.) Note further that everydayness can play a role in the scientific domain too. Biologically essentialist and supposedly scientific conceptions of gender difference involved the noted cognitive defects and had a 'common-sensical' character to them as well.[6] As with *family*, the cognitive defects tended to go unnoticed in casual observation and the conception looked fitting when it was not. The upshot is that everydayness of the distortions and cognitive errors is meaningful when distinguishing between something being merely false (*frictionless plane*) and being perniciously false (*family* traditionally conceived, biologically essentialist conception of *gender*).

Second, pernicious falsity must involve false consciousness, where beliefs are endorsed in the wrong kind of way (due to some noncognitive interests). In the case of frictionless planes, it looks like one's belief—or better, theoretical commitment—isn't epistemically warranted in the sense of being verified by observation. There are also clearly some background interests driving one's reliance on a conceptual repertoire that includes *frictionless plane*, but the interests are not noncognitive, it seems to me. In fact, I think that all scientific endeavors are driven by some prior interests, which (among other things) frame the questions asked and conceptual schemes employed. But not all such interests involve some insidious motives grounding them. What about the example of *family*? As with *frictionless plane*, relying on the traditional conception in one's conceptual scheme is not epistemically warranted. But holding on to such a scheme looks more clearly to be driven by noncognitive interests than is the case with *frictionless plane* if one is faced with many counterexamples that need to be 'explained away'. Resistance to conceptual revision, hence, does conceivably involve what Shelby calls 'false consciousness'—thereby being a component in making a concept ideologically defective.

Finally, perniciously false idealizations are ideological in having oppressive social functionality. Again, holding that a conceptual repertoire that posits *frictionless plane* establishes or reinforces unjust and oppressive social relations looks prima facie uncompelling. But idealizations that function to legitimate and

[6]A typical example of biologically deterministic view is that of Geddes and Thompson who, in 1889, argued that social, psychological, and behavioral traits were caused by metabolic state. Women supposedly conserve energy (being 'anabolic') and this makes them passive, conservative, sluggish, stable, and uninterested in politics. Men expend their surplus energy (being 'katabolic') and this makes them eager, energetic, passionate, variable and, thereby, interested in political and social matters. These biological 'facts' about metabolic states were used not only to explain behavioral differences between women and men but also to justify what our social and political arrangements ought to be. More specifically, they were used to argue for withholding from women political rights accorded to men because (according to Geddes and Thompson) "what was decided among the prehistoric Protozoa cannot be annulled by Act of Parliament" (quoted from Moi, 1999, p. 18). It would be inappropriate to grant women political rights, as they are simply not suited to have those rights; it would also be futile since women (due to their biology) would simply not be interested in exercising their political rights.

reify practices and representations through cognitively defective illusions typically have such social functionality. For instance, *family* traditionally conceived does plausibly work to reify oppressive social practices and institutions via ideological illusions.[7] Rather than seeing it as a contingent and perspectival conception, it is taken to be the (normatively and descriptively) 'standard' case to focus on with significant material consequences. One might even privilege the traditional conception of *family* as being the 'standard' in that this affords the clearest and most uncluttered view of social phenomena involving family structures. Just think back to Rawls's normative priority and distancing theses. The more complex a problem we are facing, the more simplistic and distanced must our examination of the problem be so that we can supposedly gain a clear and uncluttered view. Taking such a distanced and idealized perspective, one might hold, affords epistemic benefits in yielding a deeper understanding of the matter at hand. But as non-ideal theorists like Mills maintain, social position occupancy affects cognition, due to which apparently neutral methodological choices are deeply partial and illustrative of 'our' own social positioning. Social positions and material life circumstances shape and limit ways of knowing, understanding, and representing the world (Grasswick, 2011; Wylie, 2011). Possibilities of direct knowledge from first-hand experiences are limited because one's social position conditions the kinds of experiences that one will or can have. And so, social position occupancy shapes human cognitive perspectives and ways in which one can interpret and represent phenomena that one encounters. There need not be any explicit ill will or prejudice behind this though. It is hardly surprising that when spending time with likeminded people who all assent to some conceptual apparatus, one would come to think that there is something correct—even necessary—about ones preferred apparatus. Nonetheless, seemingly neutral conceptual choices can socially function in defective ways by reifying problematic social relations without this being driven by explicit prejudice.

What I have sketched so far is how conceptual resources may need engineering because they involve ideological defects in Shelby's sense: they involve ideological illusions, false consciousness, and oppressive social functionality. Having said that, Shelby's characterization of false consciousness is somewhat unhappy. Given what we know about the workings of human cognition, much of what we believe seemingly has its basis in "the unconscious influence of noncognitive motives" (Shelby, 2003, p. 170). If this suffices for a belief to be held irrationally, all of us most of the time are holding beliefs irrationally. Perhaps this is true; perhaps our cognitive lives are saturated with irrationality and our conceptual lives are riddled with defects. But it seems to me that irrationality and defectiveness come too cheaply on Shelby's view. Even if it is irrational to hold on to the traditional conception of *family*, *this* is not the most problematic aspect about it when considering why it is ideologically defective. To explain and clarify this, I will turn to the idea of constitutive

[7] Also think back to the example above of biologically essentialist conception of *gender*. The oppressive social functionality is strikingly obvious since it was used in arguments to withhold from women political rights.

foreclosure. This gives us a still deeper analysis of ideological defectiveness, which further puts pressure on Podosky's feasibility criterion (to be discussed in Sect. 7.6).

7.5 Constitutive Foreclosure

I take *constitutive foreclosure* to denote a phenomenon of pre-emptively excluding certain discursive and representational possibilities (see Butler, 1997; Hesse, 2014). It is not about incidentally leaving something out in the creation of meanings and representations, and hence ending up with something semantically or morally defective. Rather, constitutive foreclosure hinders and prevents the *possibility* of particular representations and renders some conceptual resources and apparatuses impossible. To illustrate, consider Mills' work elsewhere on the racial contract. He critiques the traditional social contract theory for failing to address racism and white supremacy:

> Insofar as racism is addressed at all within mainstream moral and political theory, it is usually treated in a footnote as a regrettable deviation from the ideal. But treating it this way makes it seem contingent, accidental, residual, removes it from our understanding. (1997, p. 56)

Rather, the traditional social contract establishes people of color as subpersons to be defined in opposition to free and equal citizenry to which the social contract applies. It was not an unfortunate oversight that people of color fell outside the purview of the social contract. Although race was made to seem marginal, it was as a matter of fact central to the contract and was used to establish key norms of the polity. The contract was designed to exclude people of color; hence, as Mills put it, being a racial one. Once we understand that mechanisms of exclusion are at the heart of traditional social contract theory, the idea goes, the appropriate response isn't just to insist that now people of color should be party to the contract as well. The contract itself needs to be rethought, rather than 'merely' updated.

The upshot is that ideological defects that conceptual engineers should focus on are not merely internal to the conceptual content, but also hinge on what is rendered impossible (to put it cryptically): what has been ideologically foreclosed. This does not involve 'simply' innocent failures to include some content; rather, conceptual resources that involve constitutive foreclosure presuppose that we *recognize* and *acknowledge* some states of affairs *in order to* distort them. Constitutive foreclosure is not equivalent to straightforward ignorance. Instead, it works *via* the mechanism of recognition—recognition of that which is to be distorted. This (I hold elsewhere [2022]) is an instance of the 'paradox of recognition': instances of (conceptual, theoretical, practical) misrecognition involve the recognition of some phenomenon for it to be distorted.

Think back to Mills on the racial contract. For him, it is socio-political, moral, and epistemic in prescribing cognitive norms to which (actual or would-be) signatories must adhere. In so doing, the racial contract establishes some as 'white' and coextensive with full personhood; others as 'non-white' and coextensive with subpersons who have subordinate civil standing and inferior moral status (Mills, 1997, p. 11). Hence, the social contract tradition did not innocently fail to recognize and acknowledge the importance of race. Rather, it recognized the importance of race all right, and by foreclosing race when defining the supposedly "raceless Western ideals" (Mills, 1997, p. 14), the contract was constituted in a racialized manner. This turns the social contract into "an agreement to misinterpret the world. One has to learn to see the world wrongly, but with the assurance that this set of mistaken perceptions will be validated by white epistemic authority" (Mills, 1997, p. 18). White ignorance, then, "requires a certain schedule of structured blindnesses and opacities in order to establish and maintain the white polity" (Mills, 1997, p. 19).

To clarify further, compare foreclosure to straightforward ignorance in the form of either non-recognition or misrecognition of the scope of the contract. Paul Giladi holds that misrecognition amounts to the subjectivity of a group or minority being acknowledged, but not afforded respect and value to the extent that the majority is. By contrast, non-recognition is an instance where the majority "incorrectly fails to acknowledge the subjectivity of a group or minority, affording that group or minority no positive normative status at all" (Giladi, 2018, p. 145). Constitutive foreclosure is a case of neither. The traditional contract recognized racial subjectivities precisely in order to set up a racialized polity to exclude non-white and colonized subjects (if we accept Mills' analysis). Furthermore, the traditional contract does not involve failing to see that non-white and colonized subjects share normatively relevant features with white subjects; rather, the contract took those features seriously and focused on them in defining personhood in a manner that results in a racialized normative structure. So, the issue is not about how much recognition different groups receive. It is about the *manner* in which recognition is afforded or wrestled away.

It is with this in mind that I reject Shelby's account of false consciousness in letter: ideological dirty work is being done by noncognitive motives, but these are not self-evidently unconscious. Instead, constitutive foreclosures involve motivated and willful ignorance. But this is a case of false consciousness nonetheless in the sense of being noncognitively motivated. Ignorance generated by constitutive foreclosure involves "*how* one is recognized and whether the *way* in which one is recognized is appropriate or not" (Medina, 2018, p. 3). In other words, ignorance may involve recognition, but recognition of a kind that is warped by constitutive foreclosure and that subsequently distorts some state of affairs—hence, involving ideological illusions in Shelby's sense. Ideological constitutive foreclosures recognize and acknowledge some states of affairs precisely *in order to* distort them; and these distortions have oppressive social functionality. This, I submit, accounts for the ideological nature of some conceptual defects, which isn't equivalent to 'mere' ignorance. It involves cognitive errors (distorted, biased, misleading representations), false consciousness (being influenced by noncognitive motives), and social functionality (creating and maintaining oppressive social conditions) in a manner that hinders and prevents the *possibility* of particular representations and renders some conceptual resources and apparatuses impossible.

7.6 Back to Feasibility

Given this analysis of ideological defects, what should we say about Podosky's feasibility criterion? Recall that we should engineer some ideological concept if "its desired causal influence on social reality is feasible" (Podosky, 2018, p. 13), where feasibility is about representational accuracy. We have grounds to engineer if the would-be newly engineered concepts feasibly have the causal capacity to make themselves accurate. But what if the defects are not about explicit conceptual content, instead hinging on what is foreclosed and rendered impossible to be represented? Can anything of this kind have the capacity to make themselves accurate? The idea looks highly mysterious and implausible. This is especially so since causal capacity is meant to be akin to a disposition: it would mean that something *not* found in a substance would have dispositional powers and abilities under normal stimulus conditions. If my analysis of ideological defects is on the right track, Podosky's normative feasibility criterion looks both mysterious and ineffectual. It is hard to see how any instances of conceptual engineering would be permissible if such instances cannot feasibly overcome ideological defects. Normatively speaking, then, we would have no grounds to engineer ideologically defective concepts, which strikes me as wrongheaded.

Still, a further issue pertains to de facto feasibility of engineering or implementation having accepted (on whatever grounds) that we should engineer. As I see it, my analysis of ideological defects also puts pressure on this sense of feasibility and on the implementation component of engineering. Consider once more Mills' work on white ignorance, which he takes to be a group-based cognitive hindrance. Such ignorance is not always due to 'bad faith', though it is always due to situatedness: cognitive processes are socially situated. And in white dominant and racist societies such processes are racialized in a manner that engenders white ignorance even in the absence of explicit ill will. The cognitive processes Mills discusses include perception, conception, memory, testimony, and motivational group interest (2007, pp. 23–24). Our perception is selective and socialized: despite being confronted by the 'same' object, different cognizers can quite literally perceive differently given their situatedness. Perception involves "viewing of the world through a particular conceptual grid" (Mills, 2007, p. 23), which enables us to conceive—what we see and how we interpret what we see are situated. Memory that is involved in inference from perception is also social, not merely individual. Recorded testimony is "integrated into a framework and narrative and from the start will have involved the selection of certain voices as against others, selection in and selection out" (Mills, 2007, p. 24). This being the case, in various ways interests can shape what and how we perceive, what we (individually and collectively) come to remember, whose testimony is solicited, and what putative facts and conceptual resources are accepted and reproduced.

Although not directly in relation to Mills' discussion of white ignorance, Alisa Bierria (2014) provides a fitting example of such situated cognitive processes as part of her analysis of how agency isn't dependent just on the intentions of an agent, but is also socially authored through the way others translate the agent's actions. Bierria examines the 2005-media coverage of Hurricane Katrina, and particularly two photographs published by news agencies. One image

> features white-skinned people traveling through the flooded area carrying food, and includes the caption, "Two residents wade through chest-deep water after finding bread and soda from a local grocery store after Hurricane Katrina came through the area in New Orleans, Louisiana"… The other image is nearly identical except the subject of the photo is a black man, and this time the caption reads, "A young man walks through chest deep flood water after looting a grocery store in New Orleans". (Bierria, 2014, pp. 129–130)

The same apparently intentional action is described as "finding" and "looting", respectively. Scenarios like these highlight how action is socially authored and socially read through the way others discern and translate one's action. When this takes place in systems of domination it is "facilitated by reasoning designed to reinforce and rationalize" such systems, and "further entrenches an institutionally sanctioned distortion of the intentions of some agents" (Bierria, 2014, p. 130). The looting/finding case, then, demonstrates that the social reading of agency does not involve a mere misunderstanding of the black agents' actions; rather, it involves "a fundamental corruption of the process of good faith translation" (Bierria, 2014, p. 131). This analysis fits Mills' examination of white ignorance as a cognitive hindrance that works on the levels of perception, conception, memory, testimony, and motivational group interest. All of these cognitive processes are involved in socially authored and socially read action, which in white dominant and racist societies ends up being problematically racialized.

This discussion suggests that devising conceptual resources to fix ideological defects is a much more complicated and fraught matter than conceptual engineers sometimes seemingly think. The phenomenon of constitutive foreclosure suggests that, as Mills puts it, the very "conceptual array with which the cognizer approaches the world needs itself to be scrutinized" (2007, p. 24). Implementation of engineered conceptual resources looks to require that some radical cognitive reform relative to 'our' background larger doxastic system is already underway. If 'our' cognitive processes are not undergoing radical reform, it is hard to see how engineering can be feasible. Following Podosky, normatively speaking we have no grounds to engineer ideologically defective concepts; and even if we do, this is de facto near impossible by focusing on our conceptual repertoire alone.

One might hold against this analysis of ideological defectiveness that pre-theoretically we may discover existing linguistic frameworks to be defective when they ill fit our experiences and produce jarring effects in our social existence. Rachel Sterken holds that such jarring effects may beneficially lead to a process she calls 'transformative communicative disruption':

> Changing language, while necessary, is difficult… it can often lead to miscommunication and confusion… [But] these supposed problems can actually be beneficial. It's *good* that changing language leads to miscommunication and confusion, because that can cause speakers to reflect on their language, and that will lead them to focus on its flaws and ways to improve them. (2020, p. 433)

When faced with such disruption, we can subsequently (and hopefully) find better and more apt ways to organize our conceptual worlds, where this will beneficially impact how we understand and shape ourselves as social agents, and how we can read and represent the agency of others. However, given the discussion so far, I am not optimistic about the efficaciousness of communicative disruption. Consider again the example of *family* and imagine two interlocutors who make use of different *family*-conceptions (one traditional, the other not). Let's further suppose that their exchange generates confusion due to miscommunication. How likely is it that this miscommunication will cause the interlocutors to reflect on their language? Following my analysis, it looks rather implausible that conceptual revision and innovation will ensue. Or rather, for this to happen the interlocutors would need to (be prepared to) undergo some cognitive therapy in the form of ideology critique, which is surely too much to expect from ordinary speakers in ordinary situations when faced with confusion and miscommunication. Fixing conceptual defects and engaging in conceptual engineering will be appealing to those who are already (so to speak) on board. But given how ideologies do their dirty work, getting people on board goes much beyond conceptual engineering.

To make my point stronger still, let me pause to discuss the nature of concepts in conceptual engineering, which is something that I have so far left explicitly unaddressed. My critique of Podosky and discussion up to now seemingly suggest that I focus on—what Isaac calls—the 'philosophical' conception of *concept*, whereby conceptual engineering turns "into the rather frivolous activity of fiddling with representational contents whose bearing on semantic facts is neither necessary nor sufficient to effect the desired changes in people's cognitive abilities and behaviors" (2021b, p. 11). As conceptual engineering so understood plausibly has no great impact on our cognitive lives, this speaks for a psychological approach to concepts and to conceptual engineering. Such a psychological approach takes concepts to be cognitive entities driven by "efficacy requirements":

> when concepts are activated to play a role in the cognitive processes that underwrite the exercise of our higher cognitive competences, they are intended to enable the execution of the cognitive tasks we use them for, independently of any alethic or epistemic concerns. (Isaac, 2021b, p. 10)

Such a psychological approach is prominently advanced by Edouard Machery (2021) and prima facie it looks like my critique of Podosky and analysis of ideological defects are wedded to the psychological approach. In this case though, one might wonder what (if anything) new am I suggesting in saying that conceptual engineering will be ineffective without larger changes to our doxastic architectures and without something like cognitive therapy. I agree that my view is seemingly closer to the psychological than the philosophical approach, although I do not wish to settle here on any particular conception of *concept*. But let me address how my position is related to Machery's to address the reaction that it does no new work.

Machery discusses the 'the attractor challenge' to conceptual engineering and argues that engineering prospects of some concepts are poor because they are (due to their psychological nature) attractors. On his view, "A psychological structure is an attractor if and only if (1) its influence on thought is unintentional and (2) it is more likely to determine people's trains of thought than the other psychological structures that could influence them." (2021, p. 12) Attractor concepts, then, are disposed to influence 'our' judgements about their extensions, whether we learn about extension explicitly or though cultural transmission. Hence: "Information that is consistent with the attractor is more likely to be retained; inconsistent information is more likely to be forgotten or transformed so as to fit the attractor." (Machery, 2021, p. 12) Attempts to conceptually engineer such attractor concepts are likely to be infeasible since we are (bluntly put) unlikely to feel the pull of the newly engineered concepts. As a result,

> When a folk concept is an attractor, the explicated concept is unlikely to be used in the contexts where it is supposed to be used. So, the issue is not that we cannot engineer new concepts; rather the issue is that the engineered concepts are likely to remain otiose, and that instead of the engineered concepts the lay concept will remain the effective concept. (Machery, 2021, p. 17)

According to Machery, then, it is naive to assume that a proposed reformed concept will be prioritized irrespective of the good intentions that conceptual engineers may have if they are aiming to replace lay concepts that are attractors. As examples of such lay concepts Machery notes the concept of *marriage* and *woman.* To overcome this attractor challenge, we must find "ways to counteract the influence of the lay concept on thought" (Machery, 2021, p. 20).

I take Machery's analysis of attractor concepts at face value and agree with him that *marriage* and *woman* traditionally conceived seemingly have the structure of such concepts when considering how sticky those concepts have proven to be. What I think is needed though is further analysis of why some attractor concepts are so sticky and, I contend, this is because they have ideologically defective natures. After all, not all attractor concepts look to be ideological in my sense; Machery also discusses the concept of *innateness* that looks to be a case in point. Still, to find ways to counteract the influence of ideologically defective attractor concepts like *marriage* (traditionally conceived), we need resources to understand better their ideological nature. My project here is intended to contribute to this sort of elucidation. To understand better the unintentional influence on thought and the disposition to be influenced by attractor concepts, I think that my unpacking of ideological defectiveness can do some helpful work. In short, we can examine what sorts of cognitive errors are involved in the influence that attractors have on thought (distorted, biased, misleading representations?); how being influenced by attractors involves false consciousness (how they involve being influenced by noncognitive motives), and what kind of social functionality attractors have (how they are implicated in creating and maintaining oppressive social conditions). In so doing, we can see how some conceptual resources and representational possibilities are constitutively foreclosed. Although much more should be said about this than I am able to here, my contention is that only once we understand the defectiveness of such concepts better, can we start thinking about how to affect 'our' lay conceptual schemes. And given my discussion above, it is rather likely that instead of conceptual engineering being the driving force of structural and institutional social change, the situation may be quite the opposite.

7.7 Conceptual Ethics

In sum, I am not particularly optimistic about the feasibility of engineering ideologically defective concepts that does not crucially involve a radical overhaul of our doxastic and non-doxastic lives. Or rather, at least the prospects of re-engineering and tweaking our existing conceptual resources look to be rather poor. However, let me close this paper by considering the prospects of engaging in *de novo* engineering combined with conceptual ethics. Admittedly, it is not entirely clear how Simion and Podosky understand the distinction and they do not explicitly discuss conceptual ethics: their analyses are about conceptual engineering and amelioration—but where and whether conceptual ethics comes into play, isn't clear. Nonetheless, it looks like we have ways to engage in conceptual innovation that do not involve conceptual re-engineering and tweaking. We might devise new conceptual recourses from scratch (so to speak) and then engage in the activity of conceptual ethics: in "normative and evaluative inquiry that deals with issues concerning thought and talk, such as which concepts we should use, and why" (Burgess & Plunkett, 2020, p. 287). In so doing, we can evaluate existing concepts against newly engineered ones, thus "making normative judgments as to whether they are fit for purpose" where the fittingness can involve various ends (moral, practical, alethic) (Braddon-Mitchell, 2020, p. 79; see also Burgess & Plunkett, 2013; Isaac, 2021a). Moreover, conceptual ethics is not just about *how* we should revise some linguistic resources, but also about *whether* we should employ some resources at all. If our conceptual repertoire aims to track some non-existent entities, we may have good grounds to be nihilists and no longer make use of that repertoire. Then again, even if our conceptual repertoire putatively tracks something that exists, we may have good normative grounds to stop talking in those terms and be eliminativists.

With this in mind, metalinguistic negotiation may be particularly helpful. In focusing on the pragmatics of what is being said, (normative) metalinguistic negotiation denotes a dispute "in which speakers each use (rather than mention) a term to advocate for a normative view about how that term should be used" (Plunkett, 2015, p. 832). This is not equivalent to speakers merely talking past one another, where interlocutors are not genuinely expressing a disagreement. For instance, given the national height averages, Jane who is 170 cm might be tall for a Finn but short for a Dutch. However, if Jo proclaims "Jane is tall" and Flo responds saying "Jane is not tall", we do not have a normative metalinguistic disagreement about how one ought to employ the term 'tall'—rather, for Plunkett this would be a descriptive dispute. Metalinguistic negotiations proper are not descriptive disputes, but about what a word *should* mean, or how it should be used (Plunkett, 2015, p. 838). Or as Thomasson puts it: "unlike paradigmatic verbal disputes, many cases of metalinguistic negotiation involve disputes very much worth having" (2017,

p. 12; see also Thomasson, 2020). By way of example, she notes (among others): whether the Oklahoma City bombing counts as terrorism, whether waterboarding counts as torture, and whether certain extreme performances count as art. The point here is that we are not merely engaging in disputes about the terms 'terrorism', 'torture', or 'art'. We are having *worldly*-disputes, where the interlocutors are not merely offering competing descriptions of the world, but are rather "advocating for different conceptual schemes with different impact on our way of life" (Thomasson, 2017, p. 20). Such a debate is not a superficial verbal dispute, but a substantive one. This is because normative issues about concept choices do not "depend on the way in which speakers argue about those issues. Rather, it depends on the content of those issues." (Plunkett, 2015, p. 844)

In this spirit, conceptual ethics is about advocating for different conceptual schemes and deciding among them. This is where my analysis can further do some fruitful work. One difficulty with the idea that we are 'simply' pragmatically advocating for different schemes is how to adjudicate between them—in other words, if schemes impact on our ways of life differently, which way of life should we prefer? Particularly when teaching materials of this kind, one quickly hears the complaint: "But isn't this all relative? Conceptual ethics is just pointless because we can't privilege one scheme over another!" To avoid this sort of anything-goes-relativism, advocacy of conceptual resources should precisely be subject to ideology critique: to examining if the resulting way of life that our conceptual repertoire is advocating hinges on ideological illusions, false consciousness, and has oppressive social functionality. We are subsequently still engaging in worldly-disputes—but ones that consider whether 'our' conceptual advocacy involves situated cognitive processes that give rise to ideological constitutive foreclosures. For example, imagine that my conceptual repertoire includes the non-traditional conception of *family*. In advocating for this conceptual scheme rather than the one that contains the traditional conception, I am committed to a particular way of life whereby (among other things): I recognize same-sex unions as legitimate, I am expressing my support for certain social arrangements (while rejecting others), and I engage in certain social practices that reflect my values and commitments. In considering how 'our' way of life (what are our practices, what we value, what arrangements do we support, etc) will or may be impacted by conceptual choices, we must consider whether the advocated conceptual resources are defective or distorted in ideological ways. But this task requires us to move much beyond re-engineering and tweaking of ideologically defective concepts. In other words, we may not be able to engineer without radical alterations to our cognitive architectures. Moreover, many questions that involve conceptual ethics are likely to be primarily about 'our' normative commitments rather than about conceptual fittingness, efficacy, and correctness—something that I take to be in the spirit of Machery's proposal that we must find alternative ways to undercut the influence of lay attractors on thought, as discussed above. I do not wish to conclude that we should therefore simply stop engaging in various engineering projects. But I do hold that my discussion pushes for a large dose of skepticism about the feasibility of implementation and innovation, as well as for serious modesty about how emancipatory 'our' attempts at conceptual improvement can be without this being enabled by alterations in background social conditions. Given our fallible natures, ethics is hard—and conceptual ethics is no exception.

Bibliography

Bierria, A. (2014). Missing in action: Violence, power, and discerning agency. *Hypatia, 29*(1), 129–145. https://doi.org/10.1111/hypa.12074

Braddon-Mitchell, D. (2020). Reactive concepts: Engineering the concept concept. In A. Burgess, H. Cappelen, & D. Plunkett (Eds.), *Conceptual engineering and conceptual ethics* (pp. 79–99). Oxford University Press.

Burgess, A., & Plunkett, D. (2013). Conceptual ethics I. *Philosophy Compass, 8*, 1091–1101. https://doi.org/10.1111/phc3.12086

Burgess, A., & Plunkett, D. (2020). On the relation between conceptual engineering and conceptual ethics. *Ratio, 33*(4), 281–294. https://doi.org/10.1111/rati.12265

Butler, J. (1997). *Excitable speech: A politics of the performative*. Routledge.

Cappelen, H. (2018). *Fixing language: An essay on conceptual engineering*. Oxford University Press.

Cappelen, H., & Plunkett, D. (2020). Introduction: A guided tour of conceptual engineering and conceptual ethics. In A. Burgess, H. Cappelen, & D. Plunkett (Eds.), *Conceptual engineering and conceptual ethics* (pp. 1–26). Oxford University Press.

Chalmers, D. J. (2020). What is conceptual engineering and what should it be? *Inquiry*, 1–18. https://doi.org/10.1080/0020174X.2020.1817141

Geuss, R. (1981). *The idea of a critical theory*. Cambridge University Press.

Giladi, P. (2018). Epistemic injustice: A role for recognition? *Philosophy & Social Criticism, 44*, 141–158. https://doi.org/10.1177/0191453717707237

Grasswick, H. E. (Ed.). (2011). *Feminist epistemology and philosophy of science: Power in knowledge*. Springer.

Haslanger, S., & Ásta. (2018). Feminist metaphysics. In E. Zalta (Ed.), *The Stanford encyclopedia of philosophy* (Fall 2018 ed.). https://plato.stanford.edu/archives/fall2018/entries/feminism-metaphysics/

Hesse, B. (2014). Escaping liberty: Western hegemony, blak fugitivity. *Political Theory, 42*(3), 288–313. https://doi.org/10.1177/0090591714526208

Isaac, M. G. (2021a). Post-truth conceptual engineering. *Inquiry*, 1–16. https://doi.org/10.1080/0020174X.2021.1887758

Isaac, M. G. (2021b). Which concept of concept for conceptual engineering? *Erkenntnis*, online first. https://doi.org/10.1007/s10670-021-00447-0

Isaac, M. G., Koch, S., & Nefdt, R. (2022). Conceptual engineering: A road map to practice. *Philosophy Compass, 17* (10), e12879, 1–15. https://doi.org/10.1111/phc3.12879

Machery, É. (2021). A new challenge to conceptual engineering. *Inquiry*, online first. https://doi.org/10.1080/0020174X.2021.1967190.

Medina, J. (2018). Misrecognition and epistemic injustice. *Feminist Philosophy Quarterly, 4*(4). https://doi.org/10.5206/fpq/2018.4.6233

Mikkola, M. (2022). Ideal theory, epistemologies of ignorance, and (mis)recognition. In P. Giladi & N. McMillan (Eds.), *Epistemic injustice and the philosophy of recognition* (pp. 199–216). Routledge.

Mills, C. W. (1997). *The racial contract*. Cornell University Press.

Mills, C. W. (2005). 'Ideal theory' as ideology. *Hypatia, 20*(3), 165–184. https://doi.org/10.1111/j.1527-2001.2005.tb00493.x

Mills, C. W. (2007). White ignorance. In S. Sullivan & N. Tuana (Eds.), *Race and epistemologies of ignorance* (pp. 11–38). SUNY Press.

Moi, T. (1999). *What is a woman?* Oxford University Press.
O'Neill, O. (1987). Abstraction, idealization and ideology in ethics. *Royal Institute of Philosophy Supplements, 22*, 55–69. https://doi.org/10.1017/S0957042X00003667
Plunkett, D. (2015). Which concepts should we use?: Metalinguistic negotiations and the methodology of philosophy. *Inquiry, 58*(7–8), 828–874. https://doi.org/10.1080/0020174X.2015.1080184
Podosky, P. M. C. (2018). Ideology and normativity: Constraints on conceptual engineering. *Inquiry*. https://doi.org/10.1080/0020174X.2018.1562374
Podosky, P. M. C. (2022). Don't count truth out just yet: A response to Isaac. *Inquiry*, 1–11. https://doi.org/10.1080/0020174X.2022.2075922
Rawls, J. (1971). *A theory of justice*. Harvard University Press.
Rawls, J. (1993). *Political liberalism*. Columbia University Press.
Shelby, T. (2003). Ideology, racism, and critical social theory. *The Philosophical Forum, 34*(2), 153–188. https://doi.org/10.1111/1467-9191.00132
Simion, M. (2018). The 'should' in conceptual engineering. *Inquiry, 61*(8), 914–928. https://doi.org/10.1080/0020174X.2017.1392894
Sterken, R. K. (2020). Linguistic interventions and transformative communicative disruption. In A. Burgess, H. Cappelen, & D. Plunkett (Eds.), *Conceptual engineering and conceptual ethics* (pp. 417–434). Oxford University Press.
Thomasson, A. (2017). Metaphysical disputes and metalinguistic negotiation. *Analytic Philosophy, 58*(1), 1–28. https://doi.org/10.1111/phib.12087
Thomasson, A. (2020). A pragmatic method for normative conceptual work. In A. Burgess, H. Cappelen, & D. Plunkett (Eds.), *Conceptual engineering and conceptual ethics* (pp. 435–458). Oxford University Press.
Valero, I. (2022). Ameliorating at the joints. A permissive normative framework for conceptual engineering. *Inquiry*, online first. https://doi.org/10.1080/0020174X.2022.2063174
Wylie, A. (2011). What knowers know well: Women, work and the academy. In H. Grasswick (Ed.), *Feminist epistemology and philosophy of science* (pp. 157–179). Springer.

Chapter 8
Conceptual Engineering: Rethinking "Race"

Kwame Anthony Appiah

Abstract I begin, in Part I, with four general observations about engineering that have lessons for conceptual engineering. First, thoughtful engineering requires a design specification, a rough account of what the innovations in question are aimed at. Second, once we have the specification, we have to ask what's necessary to achieve those aims. Third, there are usually many possible solutions, because (a) the aims are only vaguely specified and (b) there are all kinds of desiderata, each of which can be satisfied to a greater or lesser degree, and it's both (c) unlikely that one solution is best with respect to all the constraints and (d) not clear how to rank them. Fourth, it can be very hard to predict how a proposed innovation might be taken up. This means that very often what's needed is a trial-and-error process, in which proposals are made and we see what happens as they are and are not taken up.

In Part II, I explore three observations we need to bear in mind when it is concepts, in particular, that are being engineered. First, we need to be clear what the aim is: in practice, it is to get people to use some word (old or new) in a particular way. Since what is involved is changing the behavior of large numbers of people in a coordinated way, what is going on is, in effect, a movement for social change. Also, second, because we are working on concepts, we need to be clear about some things about concepts: (a) they are best understood as embedded in theories, so that they come in clusters not one by one, and (b) most interesting concepts have very many conceptions, perhaps, at the finest grain, as many conceptions as users of a certain word. Third, the theories in question are going to be to some degree idealized. This means, I'll argue, that truth isn't all that matters.

In Part III, I'll explore some work in reshaping understandings of race in anglophone philosophy for the last half century, with these observations in mind.

I am extremely grateful to the editors and the referees who made many useful comments on an earlier draft, and drew my attention to a number of papers that I now refer to and discuss.

K. A. Appiah (✉)
New York University, New York City, NY, USA
e-mail: Anthony.Appiah@nyu.edu; http://www.appiah.org

M. G. Isaac et al. (eds.), *New Perspectives on Conceptual Engineering - Volume 2*, Synthese Library 514, https://doi.org/10.1007/978-3-031-98524-9_8

8.1 Part I

As any engineer will tell you, you begin the design process with a specification of the task. You are not just making, say, a computer chip: you're making one that has to be able to do so and so many calculations of the right kind per second. It mustn't cost more than people will be willing to pay to use it. It should make as few errors as possible if its electric supply is stable within certain parameters and so on. Once you have a prototype up and running you can see what its capacities are, how error prone it is, and then figure out what it will cost to make many of them. How many? Well, that depends on an estimate of how many people can afford the machines it will be installed in and will prefer them to other machines with other specs.

When people say that philosophers have been doing conceptual engineering all along, I think this misunderstands the importance of the role of the design specification (Thomasson, 2021, p. 2).[1] Peter Strawson, in endorsing the interest of what he called "revisionary metaphysics"—which was not what he set out to do himself—said:

> Descriptive metaphysics is concerned to describe the actual structure of our thought about the world, revisionary metaphysics is concerned to produce a better structure. (Strawson, 1959, p. 9)

And Uriel Kriegel responded, more recently (and reasonably enough),

> Presumably what makes the structure better, however, is precisely that it captures the structure the world really has. (Kriegel, 2013, p. 1, fn. 2)

But reasonable as this is, I think it isn't right. For Kriegel is assuming that the design specification of the revisionary task is to develop concepts that more exactly represent the structure of the world. Now, I have difficulties with the idea that the world has *a* structure. But that is not my objection here. My objection is that it is always worth asking, "Better for *what*?" and not presuming that correspondence with reality is the only possible answer.

That is the main reason that philosophers have mostly not being doing conceptual engineering: they haven't usually had a well-worked-out answer to the question, "Better for what?" But, quite often, if they are assuming one, it is Kriegel's answer. That is why he was entitled to write, "Presumably …" That is what many have presumed.

Why does this matter?

Well, once you think about this design specification question, you can see that, like the engineer designing a computer chip, conceptual engineers have to think

[1] Thomasson suggests that conceptual engineering is "something we have been doing all along without realizing it," and then offers an account of conceptual engineering in terms of changing linguistic norms, about word usage, that I have much sympathy with. If you individuate concepts by their intensions and extensions, as she points out, then offering a concept with a new intension for a word like "race," is associating a word with a new concept. But she doesn't stress the importance of what I'm calling the design specification.

about what people are going to do with the thing they are designing or re-designing. (From now on, I won't keep saying "or re-designing," but, in practice, much of the conceptual engineering that is going on now is re-design: as my case study will suggest.) In this case, the object they are designing is a concept, a way of thinking about and representing the world. And in practice, to do that, they are going to propose a new way of using some words. That places constraints on what you are doing that have nothing to do with whether the concepts in question—the words in use—will represent the world (or, if you like, its structure) accurately.

Some constraints can appear trivial: if you are proposing to use an old word, say, you should think about how hard it will be to get people to forget the ways they used to use it. If you're proposing a new word, it shouldn't echo or reinforce the old ways of using words you are hoping to replace. It will help if the word trips off the tongue, and, if possible, it should evoke ideas that are usefully in the vicinity of the ones you are developing. Whether you should use an old or a new word is an important decision, and, as Cappelen has argued, you might "want to keep the old terminology, *even if it involves a change in topic*" (italics in original). One reason for doing so is associated with the fact that there are what he calls "lexical effects." "An expression," he says, "can have cognitive and emotive effects over and beyond … any of its semantic and pragmatic properties" (Cappelen, 2018, p. 123). (Examples he considers: pejoratives, the effects of the hidden slur in "niggardly," metaphor as understood by Donald Davidson.) If you think these considerations trivial, you are, I suspect, not seriously committed to bringing about conceptual change.[2] Still, that will not matter if someone else is willing to take up your ideas and do the work of shaping them for actual introduction into the world. It may be that the right people to think about that are not the conceptual engineers but the production managers, or the marketers, of the product. There is plenty of work to do to get from design specification to actual implementation: it cannot usually be done by a single person.[3]

There are usually many possible solutions to a design challenge: because, first, the aims are only vaguely specified and, second, there are all kinds of desiderata, each of which can be satisfied to a greater or lesser degree, and so, third, it's unlikely that one solution is best with respect to all the constraints and there's no agreement on how to create a single overall measure of goodness.

8.2 Part II

Other design constraints are more substantial. Having a concept expressed in a natural language entails having a disposition to use words in a particular way. So, getting a new concept into place in a language is a matter of getting people to *do* something

[2] Fischer in Fischer (2020) provides a useful discussion of some of the issues here that arise because certain conceptual structures are activated by default in every context where a word is used.

[3] Andow in Andow (2020, Sect. 5) gives a nice account of an intellectual community working together on such a project.

new. (Of course, getting them to do something new will be getting them to think in new ways, too.) That is a truth that fits with Wittgenstein's suggestion that we can sometimes answer the question "what does that mean?" by showing how a word is used. As he said in the *Investigations,*

> For a *large* class of cases—though not for all—in which we employ the word 'meaning' it can be defined thus: the meaning of a word is its use in the language. (Wittgenstein, 1953, p. 20e)

Once you have set your aim—to design a concept that will help achieve such and such ends—it's necessary, if you are to actually achieve it, that, like the engineer, you make sure it can actually be put into use. You face what the literature calls the "implementation problem."[4] And, since what is involved is changing the behavior of large numbers of people in a coordinated way, what corresponds to the production process is, in effect, a movement for social change (Bicchieri, 2016). That is true even if the conceptual re-engineering is of a concept like "gene" and your aim is not to change social relations but to change the way people think about the bodies of organisms.

A final preliminary thought. Usually when you introduce a new concept, X, you do so not by introducing a single thought, of the form "An X is a Y that F's and G's," (where Y, F and G are concepts already available), but rather by introducing a set of connected thoughts, a little theory, if you like. This is a point that Frank Ramsey brilliantly elucidated in his little essay on "Theories" (Ramsey, 1990).

Conceptual innovation, then, is often a matter of introducing not one idea but many connected ideas. You are probably familiar with the idea of a Ramsey-sentence of a theory: it's what you get when you conjoin the theory's major claims, take out a number of terms and replace them with variables, writing, for each variable "x," a quantifier "There exists an x such that," in front of the result. That you can do this reminds us that you can introduce a set of terms, a new bit of language, in one fell swoop.

Consider, by way of example, a very tidied up version of what was involved in grasping Mendel's concept of a gene.[5] There were all these new ideas:

1. Certain aspects of the phenotype of an organism are determined by its genes. (Let's call these the *genetically determined characteristics*.)
2. These genes may come in various types, called alleles, which differ in the consequences that their presence makes to the genetically determined characteristics.
3. Each of these genetically determined characteristics may exist in different forms—different colors of petals, for example, or textures of stem.

---

[4] See e.g. (Deutsch, 2020; Jorem, 2021; Koch, 2021a, b; Nimtz, 2021; Pinder, 2021; Riggs, 2019). Only Nimtz really takes seriously the fact that what's involved here is changing social norms, in the way suggested by Bicchieri 2016, though Riggs does insist that conceptual engineers "already knew they'd have to … chang[e] our social practices."

[5] I've borrowed this account from chapter four of Appiah (2003).

4. Genetically determined characteristics are produced by pairs of alleles of the gene that corresponds to them.
5. Every organism gets two alleles of each gene, one from each parent.
6. If an organism gets identical alleles from each of its parents it is homozygous for that allele, otherwise it is heterozygous.
7. If an organism is heterozygous for an allele, **A**, it has the genetically determined phenotypic characteristic corresponding to **A**, which we call the **A** phenotype.
8. An allele, **A**, must exist in one of three relations to any other allele, **A***. **A** can either

   (a) be dominant with respect to **A*** or
   (b) be recessive with respect to **A*** or
   (c) interact with **A***.

9. If **A** is dominant with respect to **A***, then an organism that is heterozygous and has the genotype **AA*** will have the **A** phenotype.
10. If **A** is recessive with respect to **A*** then an organism that is heterozygous and has the genotype **AA*** will have the **A*** phenotype.
11. If **A** interacts with **A***, then an organism that is heterozygous and has the genotype **AA***, will have neither the **A** nor the **A*** phenotype, but some other phenotype (not necessarily intermediate between **A** and **A***) that is determined by **A** and **A*** together.

    Along with these claims about how genes behave go the *laws of segregation and independent assortment of genes*.
12. **Segregation**: Each gamete bears only one of the two alleles of the adult organism.
13. **Assortment**: When two different genes separate to form gametes and join together again to form the new genotype, they do so independently.[6]

Write a Ramsey-sentence of all this, with the new words "gene," "allele," "genotype," "phenotype," "dominance," and so on, replaced by variables, and you get what was involved in grasping these new concepts, and you can see that, to get what was going on, you had to grasp them all at once.

Of course, the law of assortment is false of genes on the same chromosome, but Mendel didn't know about chromosomes. So, a conceptual alteration had to occur when they were discovered by Morgan. And that one was motivated solely by the desire to save the phenomena. It was a conceptual innovation guided by Kriegel's aim of more-accurately capturing something that you might, I suppose, loosely call the structure of the world. But, as we shall see, there are lots of other reasons for reshaping our concepts.

The fundamental thought here is what Ramsey saw as the central insight of pragmatism: beliefs are states on which we act. Or, as he more cumbrously put it, "The essence of pragmatism I take to be this, that the meaning of a sentence is to be defined by reference to the actions to which asserting it would lead, or, more vaguely

[6]This model is derived from the discussion of theories in Appiah (2003, Chap. 4).

still, by its possible causes and effects" (Ramsey, 1927, p. 170). Action is one of the central things that beliefs are for, and to be good for guiding our actions, they must be good for guiding creatures like us. We have many limitations—of memory, of computational capacity, of perceptual sensitivity—that need to be considered in designing concepts that we can actually use. That is why we have to make such extensive use of idealizations. When we introduce new concepts, in the sort of way I suggested for Mendel's gene theory, precise correspondence to reality is only one of the many things that matter. Indeed, all of our best theories are either very vague or clearly idealizations, defining "idealizations," as Hans Vaihinger did, as useful untruths (Vaihinger, 2009). As I argued in *As If: Idealizations and Ideals,* a great deal becomes much clearer if you start with Vaihinger's great insight (Appiah, 2017).

Our current best physics only works because we make certain kludges in applying it; defining a kludge as my free online dictionary does as "any system comprising several disparate parts or elements that have been poorly assembled or integrated together."[7] Relativity and the quantum theory don't naturally go together; but even if you're just trying to produce a quantum theoretic treatment of an actual system, you need to know a bunch of ad hoc tricks to get something that will make a useful prediction. Rational choice theories in the social sciences often require a model of rationality that is not just unrealized by any actual person, but actually unrealizable, since it requires logical omniscience. There are two theories of the chemical bond in current use, and nobody knows how to relate them to one another.

Once you see that, you will also see why our best chance of understanding the world must be to have a plurality of ways of thinking about it. The argument is simple. Trying to make all our theories consistent seems to make sense because the world is consistent. That was why the positivists had a program for the unification of the sciences by way of reduction. But that is too quick an argument. For if our best theories are not strictly speaking true, then the consistency of the world is not a reason for making theories consistent with one another. We need a multitude of pictures of the world, all not precisely true: and which untruths don't matter will depend on what we are using a particular theory to do. Once you see that falsehood is not by itself a reason for rejecting a theory, we can recognize the usefulness of idealizations. And once we do that, we can see that it is always worth asking about a concept or a theory, useful for what?

One reason that conceptual engineering is often re-design depends on recognizing that much of the philosophy of language I grew up with—debates about theories of meaning between realists like Davidson and anti-realists like Dummett, for example—began with a mistake. They thought that words and sentences had something called a meaning. Here is Dummett: "a theory of meaning is a theory of understanding; that is, what a theory of meaning has to give an account of is what it is that someone knows when he knows the language, that is, when he knows the meanings of the expressions and sentences of the language" (Dummett, 1996, p. 3). Here Davidson: "stated simply: a theory of meaning for a language L shows 'how the

[7] https://idioms.thefreedictionary.com/kludge

meanings of sentences depend on the meanings of words' if it contains a (recursive) definition of truth-in-L" (Davidson, 1967, p. 310).

What this all dodges is something Paul Grice adverted to in his famous paper on "Meaning," written before these two passages from Dummett and Davidson. At one point, he wrote "'x means_{NN} (timeless) that so-and-so' might as a first shot be equated with some statement or disjunction of statements about what "people" (vague) intend … to effect by x (Grice, 1957, p. 385) (The subscript there was NN for non-natural meaning, the kind of conventional meaning we see in natural languages.) And the key word here is "vague."

Maybe you can sometimes pin down the specific concept that each speaker associates with a word: to decide that you need an account of what a concept is. I myself doubt this in lots of cases because, on any plausible account of concepts, a lot of our usage requires what Hilary Putnam viewed as a linguistic division of labor (Putnam, 1975, p. 228). We mean to be referring to what the experts are referring to, when we use words like "gene" and "proton" and "elm." And deference is to a community not an individual and the individuals in that community may have different ideas.

Even ignoring that problem, it just seems to me that, as speakers develop an understanding of the language they grow into as children, there need be no precisely bounded singular concept—on whatever your favorite account is of concepts—that they all settle on. A lot of our usage has presuppositions we haven't noticed and so how we'll go on in circumstances where they fail is just not fixed in advance. What does "up" mean in a rocket circling the earth? Is a DNA sequence that provided the point of attachment for mRNA a gene?

Furthermore, while speakers in a community can regard each other as authorities on what words mean—so I can discover I have been misusing a word if most people correct my usage—there is no agreed process for settling things when usages diverge. So, there is no answer in advance to the question how to settle different alignments of words to concepts. It is a social negotiation. It is a kind of politics. And because it can be very hard to predict how a proposed innovation might be taken up, what's needed is often a trial-and-error process, in which proposals are made and we see what happens as they are and are not taken up. As Manuel Gustavo Isaac, Steffen Kock, and Ryan Nefdt have pointed out, this process raises many ethical issues: in particular, as with any engineering project, well-intentioned proposals can have unintended harmful consequences (Isaac et al., 2022).

8.3 Part III

The story so far is this: I argued, first, that conceptual engineering requires a rough account of what the innovations in question are aimed at. Once we have the specification, we have to ask what's necessary to achieve those aims. There are usually many possible solutions, because (a) the aims are only vaguely specified and (b) there are all kinds of desiderata, each of which can be satisfied to a greater or lesser degree, and it's both (c) unlikely that one solution is best with respect to all the constraints and (d) not clear how to rank them. Then, I argued that in conceptual engineering the design task is to change linguistic and other behavior. Since what is involved is changing the behavior of large numbers of people in a coordinated way, what is going on is, in effect, a movement for social change. And, because we are working on concepts, we need to be clear that they are best understood as embedded in theories, so that they come in clusters not one by one; and that most interesting concepts have very many conceptions, perhaps as many conceptions as users of a certain word. This will happen with the uptake of our newly engineered concept: it will be taken up in a variety of different ways, which you won't be able to anticipate in advance.

These are my conclusions, after thinking about the race concept, in particular, for about four decades; and thinking, as well, about the concept of social identity for a quarter century. The reason I began these explorations was somewhat contingent: I had been hired with a joint appointment in a Department of African-American and African Studies and a Department of Philosophy, and since no-one in any university, so far as I knew, had ever had such a job, it was up to me to decide what interesting work lay at the intersection of these two academic fields.

The job was in America, a country I was arriving in for the first time. I knew, as any educated modern person would, that race relations in America had been troubled from the country's start, given the role of a race concept in the definition of legal enslavement. I had by then spent much of my childhood in Africa and most of my short adulthood in Europe, and given that background, I was more familiar with the practices of race in Ghana and in Olde England than with what I discovered were the somewhat different practices in New England, where I came to work. Like many educated Ghanaians, I was aware of the writings of Richard Wright and James Baldwin and W. E. B. Du Bois, and the speeches of the Reverend Dr. Martin Luther King. (My father actually knew both Wright and Du Bois.) I had a vague sense, as someone who started following the news in my teens in the 1960s, of the American Civil Rights Movement. And I had been taught something in my high school history classes about the role of racism in the genocidal projects of Nazism. Finally—and this was another contingency—I had spent my first year at University doing Medical Sciences, since, before I realized I could do philosophy in college, I had been preparing to be a doctor. I knew more biology than most of my contemporaries.

So, I was somewhat surprised to discover that many Americans thought that race was a central biological concept. In fact, of course, in part in response to the Nazi abuses of science, there had been a couple of UNESCO statement on race that aimed to deny that the concept of race at use in social life belonged in proper scientific discourse. Looking into some of the intellectual history here, I learned that much of the work of modern cultural anthropology, beginning in the early twentieth century, had been to displace an older way of thinking about human difference in biological terms with a new one rooted in notions of cultural difference. Mendelian genetics, which really begins in the twentieth century when Morgan re-discovered and developed Mendel's hitherto largely ignored nineteenth-century work, gave an account of human difference in individual terms, and saw group differences as statistical rather than based in the racial essences of the nineteenth century.

If I had had the concept of conceptual engineering, I would have said to myself, something like this:

> Many Americans use the word "race" in ways that presuppose things that are inconsistent with the best science of our day. When it's used to refer to Black and White and Red and Yellow and Brown Americans—Negroes, Caucasians, American Indians, East, and South Asians—or to distinguish Jewish people or Hispanics in the United States, it's picking out groups that are thought to have some deep and important biological properties in common, which they don't. Of course, there are real social facts captured by assigning people to these supposed racial groups, facts that mean that there are rough-and-ready sociological generalizations that distinguish them. Blacks are poorer, on average, that Whites. They are far more likely to have suffered significant disadvantage in virtue of being regarded as Black. The same is true of American Indians (Native Americans, as some people prefer to say). Jews continue to suffer some discrimination, too, and it is not the result of their religion, since anti-Semitism is largely indifferent to people's actual religious beliefs and practices. Since the most salient fact about the race idea is that it plays a central role in communicating false beliefs—beliefs that are false by scientific standards—and in sustaining racial discrimination, surely we are better off without it. That doesn't mean we can or should ignore the reality of those sociological generalizations. But we don't need the word "race" for those groups; we can talk instead about racialized groups or about racial identities. The design specification here is this admittedly vague one: find a way to conduct everyday talk about the groups that are now called "races" that reduces both the intellectual and the moral costs of current usage.

Instead, in the absence of the idea of conceptual engineering, I wrote in 1985:

> The truth is that there are no races: there is nothing in the world that can do all we ask race to do for us. The evil that is done is done by the concept, and by easy—yet impossible—assumptions as to its application. (Appiah, 1993, p. 45)[8]

I urged, then, in effect, that we stopped "speaking of race," because there were no social groups that met the conditions that people who spoke of races took to be central to their understanding of them.

It was unsurprising, in retrospect, despite the care with which I thought I had developed this thought, that there were a variety of critical responses to racial eliminativism of this sort. I will discuss three, each of them is usefully illuminated by thinking about the issue as one conceptual design.

Before proceeding, though, let me first say a brief word about why I call this position "eliminativist." It wasn't a word I then used myself in this context, but my discussion is often treated as a locus classicus of modern eliminativism. My central eliminativist claim was that there are no races. This was not centrally a view about words. It was a view about reality, about what is out there in the world. Nothing out there in the world, I was arguing, meets the conditions that our shared understanding of the proper use of the word "race" entails. That made it a good idea not to use the word full-throatedly, so to speak: I often followed the practice of placing it in square quotes, to indicate that the speaker wasn't committed to that existing shared understanding.

[8] I made this argument first in Appiah (1985).

But once you've made it clear you think a standard way of using a term is presupposes untruths, nothing requires you to stop using the word, or to eliminate it from your vocabulary. You could think that there were better ways of using that word. So, my quick move from "no races" to "let's not speak of races" was too quick. I say this because my current view has affinities with a proposal that Daniel Wodak has made recently: that we should see eliminativism as the view that "certain uses of typical race terms are typically [morally] wrong" (Wodak, 2022, p. 587). Unlike him, I think that the harms done by much race talk are intellectual and not just moral. But we can agree that the issue isn't centrally about eliminating a word from our vocabulary.

Back, now, to those three responses. From one direction there were those who argued that abandoning the word "race" left you without a way to capture the sociological generalizations I have just been discussing. If races aren't real, how can there be obvious truths about them? My response to this line of thought was first to withdraw the argument for eliminating the term but to argue for a new understanding of it, following a proposal that W. E. B. Du Bois had made at the beginning of the twentieth century, as not a biological but a social concept, and to speak of races as socially constructed. Writing in 1911, (Appiah, 1992) Du Bois first put his point in eliminativist terms by saying that we should "speak of civilizations where we now speak of races." But he went on talking about races, for more than half a century, understanding the idea in a new way, as produced by social practices of treating people with certain purported ancestries or appearances in particular ways and regarding them as bearers of cultural traditions—which was the point of "speaking of civilizations" (Du Bois, 1911, p. 158).

Many contemporary philosophers have sought, in some such way, to ground the reference of racial terms in sociological rather than biological realities.[9] My proposal here was not one of those. My talk about racial identities goes with an account of social identities more generally that I have defended in a couple of books (Appiah, 2005, 2018). That account is nominalist, it focusses on the way in which our use of racial labels figures in way race concepts shape our self-conceptions and our treatment by others. In this way it corresponds with the proposal made by Khalifa and Lauer for a position they call the

> Constructive Antirealist Hypothesis: Some groups are racialized, but races do not exist. (Khalifa & Lauer, 2021)

The account of racialization they give comes from Sundstrom, who takes it to have three component "forces":

> …the force from above, the act of classification or labeling by some authority; the force from below, the intentional acting under the label by the so labeled; and the lateral forces, the normative standards that become attached to the label and are applied to those within and without the label. (Sundstrom, 2002a, p. 95)

[9] This section responds to some suggestions made by an anonymous reviewer.

So, accounts in terms of racialization share my focus on labeling.[10] But Sundstrom rejects nominalism about races, arguing against it in substantial detail (Sundstrom, 2002b), because he thinks that socially-created races do exist. Here he aligns with other social-scientific realists, such as Kincaid, Kelly, Mallon, Root and Wiegman (Kincaid, 2018; Mallon, 2006, 2017, 2018; Mallon & Kelly, 2012; Root, 2000; Wiegman & Mallon, 2017). What's at stake here, as Khalifa and Lauer argue, is whether the right account of the evident success of talk of races in the social sciences appeals to causal truths about races or to causal truths about how people have been racially represented. Vindicating talk of "races" in the former way requires the existence of races, vindicating it in the latter way does not. And I agree with Khalifa and Lauer that there's no reason to think that current scientific theories that use race as a variable require anything more than the Constructive Antirealist Hypothesis.

Conceptual engineers, like all engineers, have to take account of the facts, including the facts about what's really there. (To do this you need to fix on an account of what conception of X's you're relying on when you say that X's exist, of course.) I continue to think that the races of the social sciences are not really causally there: the best account of why those social science generalizations are true, I think, doesn't involve races. But that leaves me free to grant that assigning people racial labels in societies whose structure is shaped by racialization, is a useful practice. Unlike the realists, though, I don't need an answer to the question what anyone *really is* on my conception. So, when my half Norwegian, quarter Ghanaian, quarter British nephew, born in Botswana and raised mostly in Southern Africa, had a debate with my African-American god-daughters when they were all early teenagers, about whether he, with his blond hair and fair skin, was Black, unlike the realists I could retreat to discussing how various people would label him without insisting that any particular answer is right. The point is not that the question is vague: the point is that once you see what's going on, you should lose interest in the question.

Now, I didn't share Du Bois's way of thinking about socially constructed races because I never thought of them as defined by socially transmitted cultural traditions. That was the major burden of my first published paper about Du Bois (Appiah, 1985). Du Bois thought you could identify Black people by Black culture; I thought that you could only identify Black culture as the culture of Black people, and that therefore you had to have something other than culture to do the work of defining the race. But more important than this disagreement was the underlying agreement that we needed a socio-cultural story about how the not-so-clearly-bounded groups people called "races" were defined.

I should also have made more in the first paper of a point that I underlined later in Appiah (1996), namely that it wasn't just that you couldn't define races as groups of shared culture; racial groups are in fact enormously culturally inhomogeneous. Du Bois's picture of races "striving together for the accomplishment of certain more of less vividly conceived ideals of life" or "striving, each in its own way, to develop for civilization its particular message," (Du Bois, 1897, p. 76, 78) presupposes a

[10] The best account of racialization I know of is in Hochman (2017).

picture that is just inconsistent with the actual cultural lives of any of the groups we think of as races. They are not collective agents; they do not have the shared socially-transmitted ideas and practices that would make it sensible to speak of them having shared "ideals of life." So, despite Chike Jeffers' careful and interesting attempts to draw out a cultural account of races (Jeffers, 2013, 2019), I remain unconvinced.

But, apropos of Du Bois, there was a second, different, line of objection that would have come from him to much of what I was suggesting. Du Bois was the most intellectually significant of the founders of Pan-Africanism. This was a movement of solidarity about Black people, Negroes as he would have called them; despite the Africa in Pan-Africanism, the movement was centrally about bringing together people in Africa and her Atlantic Diaspora in the Americas and (to a lesser extent) in Europe. This form of solidarity was a great positive force in resisting the two central effects of racism on Black people: colonial exploitation in Africa and racial oppression elsewhere. To abandon the very word "race," some argued, was to abandon the possibility of building racial solidarity in resistance to racism.[11]

Here again, the language of conceptual engineering would have been useful. For with this design specification in mind—find ways of talking that enable anti-racist solidarity in the African Diaspora—I could have defended my own proposals as defining a Black identity for this purpose. The word "race" with its suggestions of a solidarity grounded in a natural unity rooted in a shared biology was not helpful here.

The challenge was one I identified in the next phase of the development of my work. As I wrote in 1993:

> The problem, of course, is that group identity seems to work only—or, at least, to work best—when it is seen by its members as natural, as "real." Pan-Africanism, black solidarity, can be an important force with real political benefits, but it doesn't work without its attendant mystifications. ... Recognizing the constructedness of the history of identities has seemed to many incompatible with taking these new identities with the seriousness they have for those who invent—or, as they would no doubt rather say, discover—and possess them. In sum, the demands of agency seem always—in the real world of politics—to entail a misrecognition of its genesis; you cannot build alliances without mystifications and mythologies. (Appiah, 1993, p. 175)

I argued, in response to this challenge, that the rejection of the old way of thinking about race was helpful rather than a hindrance. It allowed, for example, a distinction between two rather different projects:

> If there is, as I have suggested, hope, too, for the Pan-Africanism of an African diaspora once it, too, is released from bondage to racial ideologies (alongside the many bases of alliance available to Africa's peoples in their political and cultural struggles), it is crucial that we recognize the independence, once "Negro" nationalism is gone, of the Pan-Africanism of the diaspora and the Pan-Africanism of the continent. (Appiah, 1993, p. 180)

[11] This case was made effectively by Lucius Outlaw in Outlaw (1996). See also (Gooding-Williams, 1996) and (Jeffers, 2013).

And it made it clear since a Pan-African identity was not grounded in a natural unity, it was something that had to be built not assumed: it was, as I once argued for national identity, not a fate but a choice.

A third line of objection has come more recently from those who see in the new genomics a way of reinstalling a biological notion of race for scientific purposes. People often mention here a famous paper by Noah Rosenberg and his colleagues that used an algorithm called *structure* to partition human beings "using genotypes at 377 autosomal microsatellite loci in 1056 individuals from 52 populations." It turned out that the best clustering of these people was into 5 groups, which corresponded roughly to African, Amerindian, East Asian, European, and Melanesian origins. But, as they pointed out: "Within-population differences among individuals account for 93 to 95% of genetic variation; differences among major groups constitute only 3 to 5%" (Rosenberg et al., 2002, p. 298). So, even if there is a biological basis for this system of classification, and even if most people fit clearly into one of the five groups identified by programs like *structure,* there's a question about why *we* should be interested in this fact outside biology, given that by far the largest source of biological variation between people is within populations not across them. The point of the race concept, originally in everyday life, was to be usable to help us tell interesting things about the different groups. But if most of the interesting differences between people are within populations and aren't biological, maybe we don't need the concept of race outside biology.

Again, so it seems to me, the conceptual engineer's focus on the question "better for what," helps us to see what's what here.

I have one more point to make about the conceptual engineering point of view's utility in thinking about how to redesign our talk about what we now call "race."

As Susan Gelman argued in her fascinating book *The Essential Child,* evidence from developmental psychology shows that by the age of six children everywhere treat races as "possessing inborn features, inherent in the ... person, and passed down from parent to child." (Gelman, 2003, p. 105). Young children, she argues, also *essentialize* these groups: they believe that the "outer" characteristics by which they assign people to groups reflect shared "inner" properties that explain both appearance and behavior. It follows that any terms that we introduce into our language for kinds of people are likely to get essentialized. Of course, we can undo these habits by unteaching children these things they do naturally and spontaneously. What Sarah-Jane Leslie has called our default mode of generalization can be undone with effort. But to the extent that we think it useful for some human purposes to have rough-and-ready ways of classifying people, this psychological essentialism poses a design problem. For one of the features of race-talk that I have objected to is that it makes people expect those of one race to be more like each other—and less like members of other races—than they actually are for most important human purposes.[12]

[12] Sarah-Jane Leslie has drawn attention to the central role of generics in encouraging essentialist thinking, in e.g. (Leslie, 2017). Ritchie, in Ritchie (2021), shows how the problem is much more general.

Simply put, essentialism of this sort will often lead you astray epistemically, not just morally; but it does lead us astray morally as well because of our tendency to use kinds of people as the basis for in-group versus out-group hostility. These are costs of "race," even understood, as I am now inclined to think of it for social purposes, as a racial identity, according to an account of identity that I've sketched most recently in my book *The Lies that Bind: Rethinking Identity* (Appiah, 2018). The conceptual engineering of race as a kind of racial identity is useful for understanding how race works in our ethical lives, as I argued in *The Ethics of Identity,* earlier (Appiah, 2005). It's explicit location of race among identities, like nationality, reduces the temptation to think of the categories as made by science rather than history. But it carries the risk of essentialism that comes with any kind of identity term. It has costs as well as benefits.

As always, then, against the costs of "race" talk we must array any potential benefits. Engineering is about costs and benefits in a specific range of contexts. That is what it is to think of this as a design problem. Let me finish Part III, then, by briefly considering why I am skeptical about the value one of the most widely proposed benefits of keeping on. Doctors, enthusiasts for racial categories will tell you, need to classify people into groups of common ancestry to know what to look for. Take, they will say, sickle-cell disease and Black people. Exploring examples like this is the sort of thing you have to do in order to tally costs and benefits.

So, let's take sickle-cell disease. A normal hemoglobin molecule is made up of four subunits, 2 α and 2 β chains; each chain is produced by a distinct gene and there are many variants of both the α and the β chains. Since the α and β chains are required in equal numbers to form normal hemoglobin, there is also a range of genetic diseases associated with non-standard hemoglobins—the thalassemias—in which one or other chain is produced in too small a quantity. 39% of Nigerians have some form of α-thalassemia; the diseases produced when you have an underproduction of α chains. These disorders—the thalassemias—can be inherited both separately and together, producing a dazzling array of blood diseases, and so there is a very wide range of clinical contexts in which it is relevant to know if someone has Nigerian ancestry. Why are these genes common in Nigeria? Because it's a place where there's a lot of malaria and some of these variations in hemoglobin help you adapt to malaria. That's why they're common in the Mediterranean and the South Pacific as well, though not in the New Guinea Highlands, where the altitude makes malaria infrequent.

Of course, it's the differences in frequency between populations that make these correlations significant. As a standard discussion of blood diseases points out:

> α-Thalassemia is perhaps the most common single-gene disorder in the world. The frequency of α-thalassemia alleles is 5–10% in persons from the Mediterranean basin, 20–30% in portions of West Africa, and as high as 68% in the southwest Pacific. The frequency of heterozygote carrier status among the Chinese population has been reported to vary from 5–15%. The frequency of α-thalassemia is less than 0.01% in Great Britain, Iceland, and Japan. (Cherva et al., 2020)

Now, 25% of the population of Nigeria as a whole carries the gene for Hemoglobin S, which produces a non-standard β-chain, that leads to the classic and serious form of sickle-cell disease in homozygotes (Akinyanju, 1989). But the frequency of sickle-cell trait in the African-American population is around 1 in 13, or between 7% and 8%, because many of the ancestors of the African-American population came from non-malarious areas of Africa (and, I should remind you, Europe, Asia and the Americas, since many African-Americans have known ancestry in more than one continent in the recent past). And while the trait is carried by a lot of African-Americans, the heterozygous form occurs in only about 1 in 365 African-American births. But sickling is found in 2% of Sicilians (and is as high a 13% in some villages in Sicily): and it's found in Hispanic, South Asian, and Middle Eastern groups as well. Given that there's now a simple test for HbS that's given to every child in America at birth, it's not a good example of a characteristic for which racial information is needed in the United States.

The conceptual engineering point of view explains why this is relevant to thinking about whether, when and how to use the word "race." Its use in this case may lead medical professionals to miss the disease in, for example, people of Sicilian ancestry. And it doesn't cost much to test everybody with the relevant symptoms.

There's a more general point here. As it gets cheaper to do genomic analysis of individuals, it gets cheaper for doctors to look directly at whether someone carries a gene, rather than guessing at probabilities of carrying it on the basis of purported "race" or ancestry. If keeping track of "race" has bad effects and the good effects of knowing about genes can be achieved more reliably in another cheaply available way, the case for keeping track of "race" gets weaker. Still, because racial categories have social effects and social effects have medical effects, there will continue to be correlations between racial identities and medical facts. Social identities affect access to health care, by way of income and by way of practices of social exclusion; they affect nutrition by way of diet, through cultural transmission of dietary practices; they affect hormone levels by way of exposure to stress. The statistical distribution of genetic characteristics may account for some of the variance in health care outcomes, of course. But attention to those—which is the likely result of a focus on the biology of "race"—distracts from the social etiology of much of the variance; and thus keeps our focus away from useful approaches to reducing disparities, raising the life-chances of Black Americans by reducing the persistent effects of racism.

David Ludwig makes a useful distinction between two kinds of aims we might have in shaping a conception of race: improving public understanding and deepening our understanding of social reality. As he points out "biological and cultural specifications of 'race' are prone to producing public misunderstandings by foregrounding correlates rather than causes of racial inequality" (Ludwig, 2020, p. 492).[13] And so he argues that we should give a political constructionist account of race because that will reduce the likelihood that regular people will miss the political causes of racial differences in social outcomes. I agree with him that this, like the misdiagnoses that may result from racialized thinking in medicine, is a substantial design consideration.

[13] I'm grateful to an anonymous reviewer for the reference to Ludwig.

8.4 Coda

Some years ago, I was asked to write 100 words on my conception of philosophy, my answer was a poem that began like this.

> Let me be clear. Thinking things through is tough:
> You need the proper concepts, rightly understood.
> And no one working on their own will be enough.
> It takes a crowd of minds, working together.

I didn't then realize that the reason I thought all this was that, just as Molière's bourgeois gentilhomme was speaking prose long before his tutor drew the fact to his attention, I have been doing philosophy of race as a kind of conceptual engineering all along.[14] But recognizing that is a big step forward: it allows you an explanation of one of the central features of the debates in the philosophy of race. People are aiming to do different things when they think of as mapping the race concept. They have different implicit desiderata. That encourages pluralism about proposals, a pluralism I have long endorsed. Others have made pluralist suggestions over the years; see, recently (Mallon, 2022). My pluralism arises out of a fundamentally pragmatist conception of representational language combined with the recognition of the inevitability of idealization: we need pictures of the world to negotiate it, they're all imperfect, so we should try to improve them. But we can only do that if we realize that what counts as an improvement depends on your aims. Trying to make desiderata explicit can allow us to see that different proposals have different virtues; they achieve a variety of aims to differing degrees. And while it is hard to tell, with any project of social change, what is actually going to work, we should be guided by our best understanding of the relevant psychology and social science. If you share all the aims, you can think that, in some contexts, one notion is useful, in others, others. But this does not mean that anything goes. Some ways of doing things really are worse, all things considered, than others for certain purposes. There are better and worse conceptual engineerings of race.

References

Akinyanju, A. A. (1989). A profile of sickle cell disease in Nigeria. *Annals of the New York Academy of Sciences, 565*(1), 126–136.

Andow, J. (2020). Fully experimental conceptual engineering. *Inquiry*. https://doi.org/10.1080/0020174X.2020.1850339. Online first.

Appiah, K. A. (1985). The uncompleted argument: Du Bois and the illusion of race. *Critical Inquiry, 12*(1), 21–37.

[14] "Par ma foi! il y a plus de quarante ans que je dis de la prose sans que j'en susse rien, et je vous suis le plus obligé du monde de m'avoir appris cela." *Le bourgeois gentilhomme Act II Scene 4.*

Appiah, K. A. (1992). Social forces, 'natural' kinds. In A. Zegeye, L. Harris, & J. Maxted (Eds.), *Exploitation and exclusion: Race and class in contemporary US society* (African Discourse series 3) (pp. 1–13). Hans Zell.

Appiah, K. A. (1993). *In my father's house*. Oxford University Press.

Appiah, K. A. (1996). Race, culture, identity: Misunderstood connections. In A. Appiah & A. Gutmann (Eds.), *Color conscious: The political morality of race*. Princeton University Press.

Appiah, K. A. (2003). *Thinking it through: An introduction to contemporary philosophy*. Oxford University Press.

Appiah, K. A. (2005). *The ethics of identity*. Princeton University Press.

Appiah, K. A. (2017). *As if: Idealizations and ideals*. Harvard University Press.

Appiah, K. A. (2018). *The lies that bind: Rethinking identity*. W. W. Norton.

Bicchieri, C. (2016). *Norms in the wild: How to diagnose, measure, and change social norms*. Oxford University Press.

Cappelen, H. (2018). *Fixing language: An essay on conceptual engineering*. Oxford University Press.

Cherva, A. C., et al. (2020). What is the global prevalence of alpha thalassemia syndromes? *Medscape*. https://www.medscape.com/answers/955496-156159/what-is-the-global-prevalence-of-alpha-thalassemia-syndromes. Accessed 30 Apr 2023.

Davidson, D. (1967). Truth and meaning. *Synthese, 17*(3), 304–323.

Deutsch, M. (2020). Speaker's reference, stipulation, and a dilemma for conceptual engineers. *Philosophical Studies, 177*(12), 3935–3957. https://doi.org/10.1007/s11098-020-01416-z

Du Bois, W. E. B. (1897). The conservation of races. *American Negro Academy Occasional Papers*, No. 2, 1897, reprinted in *W.E.B. Du Bois speaks. Speeches and addresses 1890–1919* Philip S. Foner, Ed.) Pathfinders Press, 1970, 73–85.

Du Bois, W. E. B. (1911). Races. *The Crisis, 2*(4), 157–158.

Dummett, M. A. E. (1996). What is a theory of meaning? (I) In The seas of language. Oxford University Press, 1–35.

Fischer, E. (2020). Conceptual control: On the feasibility of conceptual engineering. *Inquiry*. https://doi.org/10.1080/0020174X.2020.1773309. Online first.

Gelman, S. (2003). *The essential child*. Oxford University Press.

Gooding-Williams, R. (1996). Outlaw, Appiah, and Du Bois's the conservation of races. In B. Bell, E. Grosholz, & J. Stewart (Eds.), *W. E. B. Du Bois on race and culture*. Routledge, Chapman and Hall.

Grice, H. P. (1957). Meaning. *Philosophical Review, 66*(3), 377–388.

Hochman, A. (2017). Replacing race: Interactive constructionism about racialized groups. *Ergo: An Open Access Journal of Philosophy, 4*, 61–92.

Isaac, M. G., Koch, S., & Nefdt, R. (2022). Conceptual engineering: A road map to practice. *Philosophy Compass, 17*(10), e12879, 1–15. https://doi.org/10.1111/phc3.12879

Jeffers, C. (2013). The cultural theory of race: Yet another look at Du Bois's 'the conservation of races'. *Ethics, 123*(3), 403–426.

Jeffers, C. (2019). Cultural Constructionism. In J. Glasgow, S. Haslanger, C. Jeffers, & Q. Spencer (Eds.), *What is race? Four philosophical views* (pp. 38–72). Oxford University Press.

Jorem, S. (2021). Conceptual engineering and the implementation problem. *Inquiry, 64*(1–2), 186–211. https://doi.org/10.1080/0020174X.2020.1809514

Khalifa, K., & Lauer, R. (2021). Do the social sciences vindicate race's reality? *Philosophers Imprint, 21*, 1–17.

Kincaid, H. (2018). Debating the reality of race, caste, and ethnicity. *Philosophy of the Social Sciences, 48*(2), 139–167.

Koch, S. (2021a). The externalist challenge to conceptual engineering. *Synthese, 198*(1), 327–348. https://doi.org/10.1007/s11229-018-02007-6

Koch, S. (2021b). There is no dilemma for conceptual engineering: Reply to Max Deutsch. *Philosophical Studies, 178*(7), 2279–2291. https://doi.org/10.1007/s11098-020-01546-4

Kriegel, U. (2013). The epistemological challenge of revisionary metaphysics. *Philosophers' Imprint, 13*(12), 1–30.

Leslie, S.-J. (2017). The original sin of cognition: Fear, prejudice, and generalization. *Journal of Philosophy, 114*, 393–421.

Ludwig, D. (2020). Understanding race: The case for political constructionism in public discourse. *Canadian Journal of Philosophy, 50*(4), 492–504.

Mallon, R. (2006). Race: Normative, not metaphysical or semantic. *Ethics, 116*(3), 525–551.

Mallon, R. (2017). Social construction and achieving reference. *Noûs, 51*(1), 113–131.

Mallon, R. (2022). What's at stake in the race debate? *Southern Journal of Philosophy, 60*(S1), 54–72.

Mallon, R., & Kelly, D. (2012). Making race out of nothing: Psychologically constrained social roles. In H. Kincaid (Ed.), *The Oxford handbook of philosophy of social science* (pp. 507–530). Oxford University Press.

Nimtz, C. (2021). Engineering concepts by engineering social norms: Solving the implementation challenge. *Inquiry*. https://doi.org/10.1080/0020174X.2021.1956368. Online first.

Outlaw, L. (1996). "Conserve" races? In defense of W.E.B. Du Bois. In B. W. Bell, E. R. Grosholz, & J. B. Stewart (Eds.), *W. E. B. Du Bois on race and culture: Philosophy, politics, and poetics* (pp. 15–38). Routledge.

Pinder, M. (2021). Conceptual engineering, metasemantic externalism and speaker-meaning. *Mind, 130*(517), 141–163. https://doi.org/10.1093/mind/fzz069

Putnam, H. (1975). The meaning of "meaning". In *Mind, language and reality* (pp. 215–271). Cambridge University Press.

Ramsey, F. P. (1927). Facts and propositions. *Proceedings of the Aristotelian Society (Supplementary Volume), 7*, 153–170.

Ramsey, F. P. (1990). Theories. In D. H. Mellor (Ed.), *Philosophical papers* (pp. 112–136). Cambridge University Press.

Riggs, J. (2019). Conceptual engineers shouldn't worry about semantic externalism. *s*, 1–22. https://doi.org/10.1080/0020174X.2019.1675534. Online first.

Ritchie, K. (2021). Essentializing language and the prospects for ameliorative projects. *Ethics, 131*(3), 460–488.

Root, M. (2000). How we divide the world. *Philosophy of Science, 67*, S628–S639.

Rosenberg, et al. (2002). Genetic structure of human populations. *Science, 298*(5602), 2381–2385. https://doi.org/10.1126/science.1078311

Strawson, P. F. (1959). *Individuals: An essay in descriptive metaphysics*. Routledge.

Sundstrom, R. R. (2002a). Race as a human kind. *Philosophy & Social Criticism, 28*(1), 91–115.

Sundstrom, Ronald R. (2002b). 'Racial' Nominalism. Journal of Social Philosophy 33 (2):193–210.

Thomasson, A. L. (2021). Conceptual engineering. When do we need it? How can we do it? *Inquiry*. https://doi.org/10.1080/0020174X.2021.2000118. Online first.

Vaihinger, H. (2009). *The philosophy of "as if"* (C. K. Ogden, Trans.). Martino Publishing.

Wiegman, I., & Mallon, R. (2017). Applied philosophy of social science: The social construction of race. In K. Lippert-Rasmussen, K. Brownlee, & D. Coady (Eds.), *A companion to applied philosophy* (pp. 441–454). Wiley Blackwell.

Wittgenstein, L. (1953). *Tractatus Logico-Philosophicus*. Routledge & Kegan Paul.

Wodak, D. (2022). Of witches and white folks. *Philosophy and Phenomenological Research, 104*(3), 587–605.

MIX
Papier aus verantwortungsvollen Quellen
Paper from responsible sources
FSC® C105338

If you have any concerns about our products,
you can contact us on
ProductSafety@springernature.com

In case Publisher is established outside the EU,
the EU authorized representative is:
Springer Nature Customer Service Center GmbH
Europaplatz 3, 69115 Heidelberg, Germany

Printed by Libri Plureos GmbH
in Hamburg, Germany